Introduction to Emergency Management

Introduction to Emergency Management

Paul Morgan

STATES
ACADEMIC PRESS
www.statesacademicpress.com

States Academic Press,
109 South 5th Street,
Brooklyn, NY 11249, USA

Visit us on the World Wide Web at:
www.statesacademicpress.com

ISBN: 978-1-63989-305-8 (Hardback)

Cataloging-in-Publication Data

Introduction to emergency management / Paul Morgan.
p. cm.
Includes bibliographical references and index.
ISBN 978-1-63989-305-8
1. Emergency management. 2. Public safety. 3. First responders.
4. Natural disasters--Management. I. Morgan, Paul.

HV551.2 .I58 2022
363.348--dc23

Table of Contents

Preface VII

Chapter 1 What is Emergency Management? 1

a. Emergency Planning Ideals 8
b. Thunderstorm and Tornado Preparedness 21
c. Food Safety in Emergency Situations 31
d. Survival Kit 40
e. Humanitarian Aid 40

Chapter 2 Disasters and their Management 42

a. Earthquake 42
b. Flood 51
c. Tsunami 58
d. Tornado 68
e. Landslide 76

Chapter 3 Preparedness and Response 85

a. Preparedness 85
b. Disaster Response 104

Chapter 4 Recovery and Mitigation 117

a. Recovery 117
b. Mitigation 144
c. Guidelines for Planning in Re-building Process 159

Chapter 5 Warning Systems 179

a. Fire Alarm System 183
b. Emergency Population Warning 191
c. Emergency Alert System 192
d. Tsunami Warning System 201

e. Earthquake Warning System 206
f. Gale Warning 209

Chapter 6 Emergency Management Software 210
a. Real-time Emergency Management Applications 211
b. Emergency Management Information System 237
c. Advanced Disaster Management Simulator 238

Permissions

Index

Preface

This book has been written, keeping in view that students want more practical information. Thus, my aim has been to make it as comprehensive as possible for the readers. I would like to extend my thanks to my family and co-workers for their knowledge, support and encouragement all along.

Emergency refers to a state that disrupts the natural procedures. It requires immediate measures to be taken in order to prevent it from converting into a disaster. Emergency management is concerned with the management and organization of resources and responsibilities during an emergency. It is primarily focused on the humanitarian aspects of handling emergencies such as preparedness, mitigation, response and recovery. Preparedness refers to preparing procedures and equipment which can be used in the event of a disaster. Proactive measures which aid in eliminating and reducing the risks as well as impacts of disasters fall under the category of disaster mitigation. The response phase of emergency management focuses on the fulfillment of the humanitarian needs of the affected population. Once the immediate threat to human life is subsided, the recovery phase is initiated. This book studies, analyses and uphold the pillars of emergency management and its utmost significance in modern times. It is a compilation of chapters that discuss the most vital concepts in this field. This textbook is an essential guide for both academicians and those who wish to pursue this discipline further.

A brief description of the chapters is provided below for further understanding:

Chapter - What is Emergency Management?

Emergency management is defined as the management of resources and responsibilities related to preparedness, response, mitigation and recovery of emergencies. It aims at reducing the harmful effects of all hazards and disasters. This chapter has been carefully written to provide an easy understanding of emergency management.

Chapter - Disasters and their Management

There are various geophysical, hydrological, climatological, meteorological and biological disasters that fall under emergency management. It includes earthquakes, landslides, folds, drought, cyclones, disease epidemics, etc. The topics elaborated in this chapter will help in gaining a better perspective of these various types of disasters and their management.

Chapter - Preparedness and Response

Preparedness involves research, estimation and planning of precautionary measures to counter the potential disasters. Disaster response includes searching and rescuing, providing immediate assistance and restoration of infrastructure for supporting life of the affected population. All the aspects related to preparedness and response in emergency management have been carefully analyzed in this chapter.

Chapter - Recovery and Mitigation

Recovery deals with the rebuilding and repair of destroyed infrastructure for restoring community's systems and activities to normal. Mitigation lessens the impact of disasters and reduces the loss of life and property. This chapter discusses the concepts associated with recovery and mitigation in emergency management in detail.

Chapter - Warning Systems

Fire alarm system, emergency population warning system, emergency alert system, tsunami warning system, earthquake warning system, gale warning system, etc. are some of the warning systems that are used in emergency management. This chapter closely examines the different warning systems used in emergency management for a thorough understanding of the subject.

Chapter - Emergency Management Software

There are many emergency management software and applications that are widely used. This chapter sheds light on these emergency management software to provide an in-depth understanding of the subject.

Paul Morgan

1

What is Emergency Management?

Emergency management is defined as the management of resources and responsibilities related to preparedness, response, mitigation and recovery of emergencies. It aims at reducing the harmful effects of all hazards and disasters. This chapter has been carefully written to provide an easy understanding of emergency management.

Emergency management is the discipline of dealing with and avoiding risks, particularly those that have catastrophic consequences for communities, regions, or entire countries. It is the dynamic process of preparing for, mitigating, responding to and recovering from an emergency. Planning, though critical, is not the only component. Training, conducting drills, testing equipment and coordinating activities with the community are other important functions. Effective emergency management relies on the integration of emergency plans at all levels of government and non-government, including individuals and community organizations.

Emergency

An emergency is any unplanned event that can cause deaths or significant injuries to the public; or that can disrupt operations, cause physical or environmental damage, or threaten the financial standing of businesses and institutions. Obviously, numerous events can be "emergencies," including:

- Wildfires,
- Hazardous materials incidents,
- Floods or flash floods,
- Hurricanes,
- Tornadoes,
- Winter storms,

- Earthquakes,
- Communications, public transportation, or electricity failures,
- Radiological accidents,
- Civil disturbances, terrorism, war,
- Loss of key suppliers,
- Explosions.

The four stages of emergency management:

Mitigation

Mitigation is taking action now—before the next disaster—to reduce human and financial consequences later. It involves analyzing risk, reducing risk, and insuring against risk. Effective mitigation requires that we all understand local risks, address the hard choices, and invest in long-term community well-being. Mitigation is achieved through regulations, local ordinances, land use and building practices.

Preparedness

Preparedness is a continuous cycle of planning, managing, organizing, training, equipping, exercising, creating, monitoring, evaluating and improving activities to ensure effective coordination and the enhancement of capabilities of concerened organizations to prevent, protect against, respond to, recover from, create resources and mitigate the effects of natural disasters, acts of terrorism, and other man-made disasters.

Response

Response includes the mobilization of the necessary emergency services and first responders. This is driven by the type and kind of emergency and is likely to include a first wave of core emergency services, such as firefighters, police and ambulance crews. They may be supported by a number of secondary emergency services. A well rehearsed emergency plan makes rescue and response more efficient.

Recovery

The aim of recovery is to restore the affected area to its previous state. It differs from response in its focus; recovery efforts deal with issues and decisions that must be made after immediate needs are met. Recovery efforts are primarily concerned with actions that involve rebuilding destroyed property, re-employment, and the repair of essential infrastructure. Efforts should be made to "build back better," with a goal to reduce risks inherent in the community and infrastructure.

Preparing a Disaster Preparedness Plan

There is no rigid or standard format for disaster preparedness plans, in fact, the reverse applies, in that the format needs to fit the circumstances in which the plan is being made and the requirements, which it is designed to meet. The essential point is that the disaster preparedness plan should be exhaustive but also practical. Hence certain common features tend to apply to most plans.

Clarity of Aim

The aim needs to be carefully and accurately selected, because it determines the whole thrust and scope of the plan. All information, guidelines, directions and instructions, which are included in the plan, need to be in line with the aim.

Realism

The plan must be realistic in the sense that it relates to an accurate assessment of the disaster threat and the vulnerability of the community and that it takes, into account the scale and capability of counter-disaster resources which are available. In most cases, the plan will be used under difficult disaster circumstances when, perhaps, communications are adversely affected; therefore, the plan should be able to respond to various contingencies during a disaster scenario.

Level of the Plan

The plan must be accurately related to the level with which it is concerned, e.g., Village level, Block level, Taluka or Tehsil level or District level.

Flexibility

Disaster circumstances tend to vary and do not necessarily follow set patterns, counter-disaster plans also need to be flexible. Flexibility is best achieved by planning to cope with the full range of possible disaster threats and ensuring that, within the overall plan, response arrangements can be rapidly adapted to new and changing circumstances. Planned decentralization, where appropriate, is a useful way towards achieving flexibility.

Coordination

Since coordination of efforts is a key factor in counter-disaster activities, the planshould include an optimum system for direction/coordination.

Assignment of Responsibility

It is critically important that responsibilities are clearly and unambiguously defined and assigned in the plans. This reduces to a minimum the possibility of misunderstandings, duplications and omissions in the various activities the plan covers. Of equal importance is the fact that clear definition of responsibilities significantly helps in achieving coordination of effort.

Ease of Use

The plan should be formulated in such a way that it is easy to understand and easy to use. References within the plan should be clear and readily identifiable. Also, the text of the plan needs to be kept as clear and concise as possible, with annexes being used for very detailed information.

Plan Components

There are a number of options for dividing the plan into sections or components. One way this can be achieved is to have:

- A main plan (or main action plan) which contains the primary parts of the plan, such as the anticipated disaster threat, vulnerability of the community (including its strengths and weaknesses in relation to each anticipated disaster scenario), the main requirements for dealing with the threat, resources, organisation, direction and coordination, warning, operational implementation of the plan, counter-disaster operations, recovery policy, and post-disaster review process.

- Sub-plans which are a part of the main plan but which may be required to amplify parts of the main plan which need special consideration, such as evacuation, relief camps, public information, and so on.

- Special plans which may be required to deal with special contingencies such as an outbreak of common or rare disease, which would require specialist personnel and procedures. Such special plans would normally be designed to work in harmony with the main plan and utilize the overall counter-disaster effort as necessary.

Viability P – 27

The Plan should include arrangements for periodically reviewing so that it is kept up-to-date and fully viable for the purposes for which it is designed.

Structure of the Plan

A typical disaster preparedness plan is structured as follows:

- Contents,
- Authorization,
- Map References,

- Introduction:
    - The Disaster Threat,
    - National Policy and State Policy,
    - General Concept for Disaster Action.
- Aim of the Plan,
- Definitions,
- Relationships with other Plans,
- Main requirements for dealing with Disasters in the area,
- Emergency Powers:
    - Disaster Legislation,
    - Other Legislation.
- Counter-Disaster Resources:
    - Within the area,
    - In the neighborhood.
- Organizational Structure and Responsibilities:
    - Prime Minister,
    - Chief Minister/Administrator (of Union Territory),
    - Nodal Ministry at the Centre and State,
    - National Disaster Management Structure,
    - State or Regional Disaster Management Committees,
    - District Disaster Management Structure,
    - Non-government Organizations active in the area and their specializations and resources,
    - Other community based organizations in the area,
    - Defence Services and Para Military Forces available in the area for Disaster Management work,
    - International Assistance Arrangements,
    - Coordination of Planning, Organizational and Operational Measures-Control Room,

  - Media cooperation.
- Preparedness Measures:
  - General,
  - National Level,
  - State Level,
  - District and Community Level,
  - Training and Public Awareness Programmes.
- Communications (General and Emergency),
- Operational Direction and Coordination:
  - Responsibility for Operational Direction and Coordination,
  - National Emergency Operations Centre – Control Room,
  - State Emergency Centres – Control Room,
  - District Emergency Centre (Control Room),
  - Field Control Rooms.
- Warning Arrangements:
  - General,
  - Agencies Originating Warnings,
  - Transmission of Warnings,
  - Dissemination and Public Broadcast of Warnings,
  - Notification of de-alert or Ail-Clear messages.
- Operational Implementation of Plan,
- Stages of Implementation,
- Counter-Disaster (or Response) Operations:
  - Precautionary Measures,
  - Activation of Emergency Operations Centres (Control Rooms),
  - Direction and Coordination of Operations,
  - Information Requirements,
  - Operational Requirements in disaster stricken areas,

    - Operational Action – National Level,
    - Operational Action – State, Regional and Local Levels,
    - Period of Disaster Operations (Emergency Phase).
- Recovery:
    - Statement of Policy for Recovery,
    - Responsibility for Recovery Programme,
    - Cross-reference to Recovery Plan (if applicable).
- Post-Disaster Review:
    - Responsibility,
    - Debriefing,
    - Review of Plans and Organizations.
- Support Measures:
    - Training,
    - Public Awareness.
- Annexures:
    - Distribution Lists,
    - Telephone Numbers, cell phone numbers and addresses of functionaries,
    - List of Resources,
    - Functional Diagram of Organisation,
    - Allocation of Roles and Responsibilities to Resource Organisation,
    - Guidelines for International Assistance Arrangements,
    - Communications,
    - Detailed Information on Warning,
    - Precautionary Measures on Receipt of Warning,
    - Guidelines on Training,
    - Guidelines on Public Awareness,
    - Format for Departmental Standard Operational Procedures,
    - List of media persons and agencies with telephone numbers.

Cyclone Preparedness

The preparedness means measures, which enable government agencies, private organizations, communities, and individuals to respond rapidly and effectively to disaster situations. The preparedness measures include the formulation of viable disaster mitigation plans.

The preparedness actions have to be planned ahead of disaster. It would consist of a plan of action to be implemented on the receipt of the Cyclone Alert message from Cyclone Warning Centre. A cyclone alert is issued generally 48 hours before the possibility of the area being affected by cyclonic weather such as strong winds, heavy rain and storm surges.

The Action Plan would indicate how evacuation of people would be effected and the places where they could be evacuated. The identification of strong buildings, which would withstand the fury of the storm, is an important segment of preparedness action plan. The safe storage of non-perishable food and other essential needs, adequate collection of stocks of drinking water and medicines, has to be made.

Most of the maritime states have prepared Cyclone Disaster Preparedness handbooks or manuals, where action plans of various organizations have been indicated in the case of cyclone threat. It is desirable-that as an essential component of preparedness, the action points indicated in the manuals are rehearsed at the beginning of each cyclone season and updated in the light of experience gathered.

Disaster Preparedness for Floods

In the context of floods, it is well known that floods damage human settlements, necessitate evacuation to safer areas, damage crops and disrupt farming, wash away infrastructure items like irrigation, communication etc. and make land unusable. Disaster preparedness should also deal with all these aspects and other connected matters.

The vary basic step in vulnerability reduction will be to identify such high risk areas, prepare risk maps showing the likely risks at different probability levels of flooding and make this knowledge available widely.

Emergency Planning Ideals

Identification of Risks

Risk is defined as the probability of harmful consequences, or expected losses (deaths, injuries, property, livelihoods, economic activity disrupted or environment damaged) resulting from interactions between natural or human-induced hazards and vulnerable conditions. Risk can presented conceptually with the following basic equation indicated in figure. Risk can presented conceptually with the following basic equation:

Risk = Hazard * Vulnerability * Amount of elements-at-risk

The equation given above is not only a conceptual one, but can also be actually calculated with spatial data in a GIS to quantify risk from geomorphological hazards. The way in which the amount of elements-at-risk are characterized (e.g. as number of buildings, number of people, economic value) also defines the way in which the risk is presented. Table gives a more in-depth explanation of the various components involved. In order to calculate the specific risk equation 1 can be modified in the following way:

$$R_S = P_{(T:Hs)} * P_{(L:Hs)} * V_{(Es \mid Hs)} * A_{ES}$$

in which:

$P_{(T:Hs)}$ is the temporal (e.g. annual) probability of occurrence of a specific hazard scenario (H_S) with a given return period in an area;

$P_{(L:Hs)}$ is the locational or spatial probability of occurrence of a specific hazard scenario with a given return period in an area impacting the elements-at-risk;

$V_{(Es \mid Hs)}$ is the physical vulnerability, specified as the degree of damage to a specific element-at-risk E_S given the local intensity caused due to the occurrence of hazard scenario H_S.

A_{Es} is the quantification of the specific type of element at risk evaluated (e.g. number of buildings).

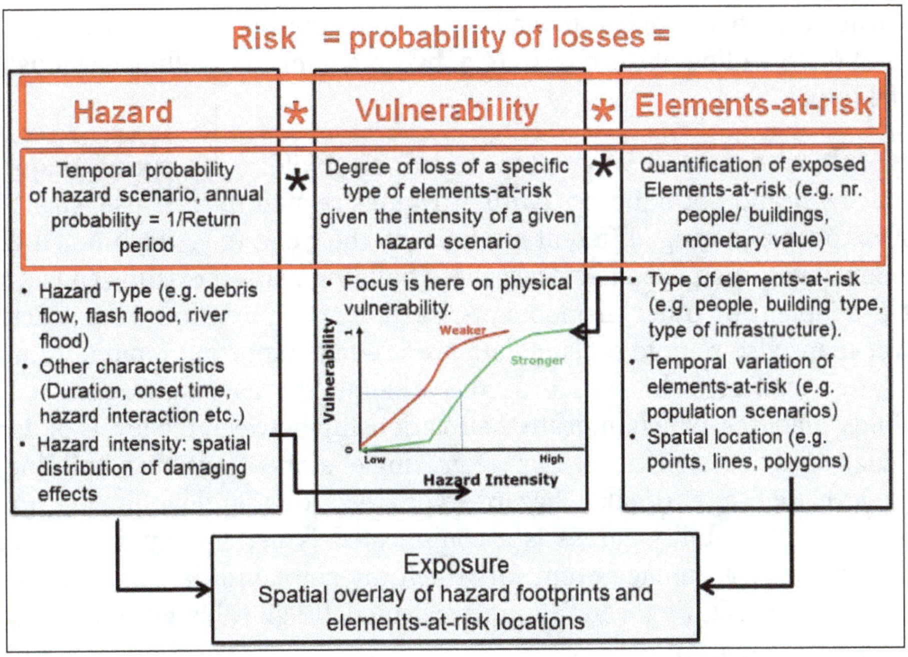

Schematic representation of risk as the multiplication of hazard, vulnerability and quantification of the exposed elements-at-risk. The various aspects of hazards, vulnerability and elements-at-risk and their interactions are also indicated. This framework focuses on the analysis of physical losses, using physical vulnerability data.

The term risk mapping is often used as being synonymous with risk analysis in the overall framework of risk management. Risk assessments (and associated risk mapping) include: A review of the technical characteristics of hazards such as their location, intensity, frequency and probability; the analysis of exposure and vulnerability including the physical social, health, economic and environmental dimensions; and the evaluation of the effectiveness of prevailing and alternative coping capacities in respect to likely risk scenarios. In the framework of natural hazards risk assessment, the term risk mapping also indicates the importance of the spatial aspects of risk assessment. All components of the risk equation are spatially varying and the risk assessment is carried out in order to express the risk within certain areas. To be able to evaluate these components we need to have spatially distributed information. Computerized systems for the collection, management, analysis and dissemination of spatial information, so-called Geographic Information Systems (GIS) are used to generate the data on the various risk components, and to analyze the risk . Hazard data is generally the most difficult to generate. For each hazard type (e.g. flooding, debrisflow, rockfall) so-called hazard scenarios should be defined, which are hazard events with a certain magnitude/intensity/frequency relationship (e.g. flood depth maps for 10, 50 and 100 year return periods). Different types of modelling approaches are required for the hazard scenario analysis, depending on the hazard type, scale of analysis, availability of input data, and availability of models. Generally speaking a separate analysis is required to determine the probability of occurrence for a given magnitude of events, followed by an analysis of the initiation of the hazard (e.g. hydrological modelling or landslide initiation modelling), and of the run-out or spreading of the hazard (e.g. hydrodynamic modelling or landslide run-out modelling).

Elements-at-risk data are very often based on building footprint maps, which represent the location of buildings, with attributes related to their use, size, type and number of people during different periods of the year (e.g. daytime, nighttime). Remote Sensing is often used to extract these building maps if existing cadastral maps are not available. For other elements-at-risk like transportation infrastructure and land cover maps also remote sensing data are used as important inputs. Vulnerability data is often collected in the form of vulnerability curves, fragility curves or vulnerability matrices, which indicate the relationship between the levels of damage to a particular type of element-at-risk (e.g. single storey masonry building) given intensity levels of a particular hazard type (e.g. debris flow impact pressure). Generation of vulnerability curves is a complicated issue, as they can be generated empirically from past damage event for which intensity and damage is available for many elements-at-risk, or through numerical modelling. Table gives an overview of the various components of risk.

Term	Definition	Equations & explanation
Natural hazard (H)	A potentially damaging physical event, phenomenon or human activity that may cause the loss of life or injury, property damage, social and economic disruption or environmental degradation. This event has a probability of occurrence within a specified period of time and within a given area, and has a given intensity.	$P_{(T:HS)}$ is the temporal (e.g. annual) probability of occurrence of a specific hazard scenario (H_s) with a given return period in an area; $P_{(L:HS)}$ is the locational or spatial probability of occurrence of a specific hazard scenario with a given return period in an area impacting the elements-at-risk
Elements-at-risk (E)	Population, properties, economic activities, including public services, or any other defined values exposed to hazards in a given area". Also referred to as "assets".	E_s is a specific type of elements-at-risk (e.g. masonry buildings of 2 floors)
Vulnerability (V)	The conditions determined by physical, social, economic and environmental factors or processes, which increase the susceptibility of a community to the impact of hazards. Can be subdivided in physical, social, economical, and environmental vulnerability.	$V_{(Es \mid Hs)}$ is the physical vulnerability, specified as the degrees of damage to E_S given the local intensity caused due to the occurrence of hazard scenario H_S It is expressed on a scale from 0 (no damage) to 1 (total loss)
Amount of elements-at-risk (A_E)	Quantification of the elements-at-risk either in numbers (of buildings, people etc), in monetary value (replacement costs etc), area or perception (importance of elements-at-risk).	A_{Es} is the quantification of the specific type of element at risk evaluated (e.g. number of buildings)
Consequence (C)	The expected losses (of which the quantification type is determined by A_E) in a given area as a result of a given hazard scenario.	C_S is the "specific consequence", or expected losses of the specific hazard scenario which is the multiplication of $V_S * A_{ES}$
Specific risk (R_S)	The expected losses in a given area and period of time (e.g. annual) for a specific set of elements-at-risk as a consequence of a specific hazard scenario with a specific return period.	$R_S = H_S * V_S * A_{ES}$ $R_S = H_S * C_S$ $R_S = P_{(T:Hs)} * P_{(L:Hs)} * V_{(Es \mid Hs)} * A_{ES}$
Total risk (R_T)	The probability of harmful consequences, or expected losses (deaths, injuries, property, livelihoods, economic activity disrupted or environment damaged) resulting from interactions between natural or human-induced hazards and vulnerable conditions in a given area and time period. It is calculated by first analyzing all specific risks. It is the integration of all specific consequences over all probabilities.	$R_T \approx \sum (R_T) = \sum (H_S * V_S * A_{ES})$ Or better: $R_T = \int (V_S * A_{ES})$ - For all hazard types - For all return periods - For all types of elements-at-risk. It is normally obtained by plotting consequences against probabilities, and constructing a risk curve. The area below the curve is the total risk.

Risk Assessment and Mapping

Risk mapping for natural hazard risk can be carried out at a number of scales and for different purposes. Table gives a summary. In the following sections four methods of risk mapping will be discussed: Quantitative risk assessment (QRA), Event-Tree Analysis (ETA), Risk matrix approach (RMA) and Indicator-based approach (IBA).

Table: Indication of scales of analysis with associated objectives and data characteristics (approaches: QRA = Quantitative risk assessment, EVA = Event-Tree Analysis, RMA = Risk matrix approach, IBA = Indicator-based approach).

Scale of analysis	Scale	Possible objectives	Possible approaches
International, Global	< 1 : 1 million	Prioritization of countries/regions; Early warning.	Simplified RMA & IBA
Small: provincial to national scale	< 1:100,000	Prioritization of regions; Analysis of triggering events; Implementation of national programs; Strategic environmental assessment; Insurance.	Simplified EVA, RMA & IBA
Medium: municipality to provincial level	1:100000 to 1:25000	Analyzing the effect of changes; Analysis of triggering events; Regional development plans.	RMA / IBA
Local: community to municipality	1:25000 to 1:5000	Land use zoning; Analyzing the effect of changes; Environmental Impact Assessments; Design of risk reduction measures.	QRA / EVA / RMA IBA
Site-specific	1:5000 or larger	Design of risk reduction measures; Early warning systems; detailed land use zoning.	QRA / EVA / RMA

The Quantitative Risk Assessment method is the best for evaluating several alternatives for risk reduction, through a comparative analysis of the risk before and after the implementation followed by a cost-benefit analysis. The event-tree analysis is the best approach for analyzing complex chains of events and the associated probabilities. The risk matrix approach is often the most practical approach as basis for spatial planning, where the effect of risk reduction methods can be seen as changes in the classes within the risk matrix. The indicator-based approach, finally, is the best when there is not enough data to carry out a quantitative analysis, but also as a follow-up of a quantitative analysis as it allows to take into account other aspects than just physical damage. Even though hazard and risk mapping may have taken place, real risk reduction will only happen when it leads to a reduction in either the hazards frequency and intensity, the number of exposed elements-at-risk and their vulnerability. This requires integration of risk analysis into a risk management framework, which includes the adoption of policy and regulations, and interaction of geoscientists within this process.

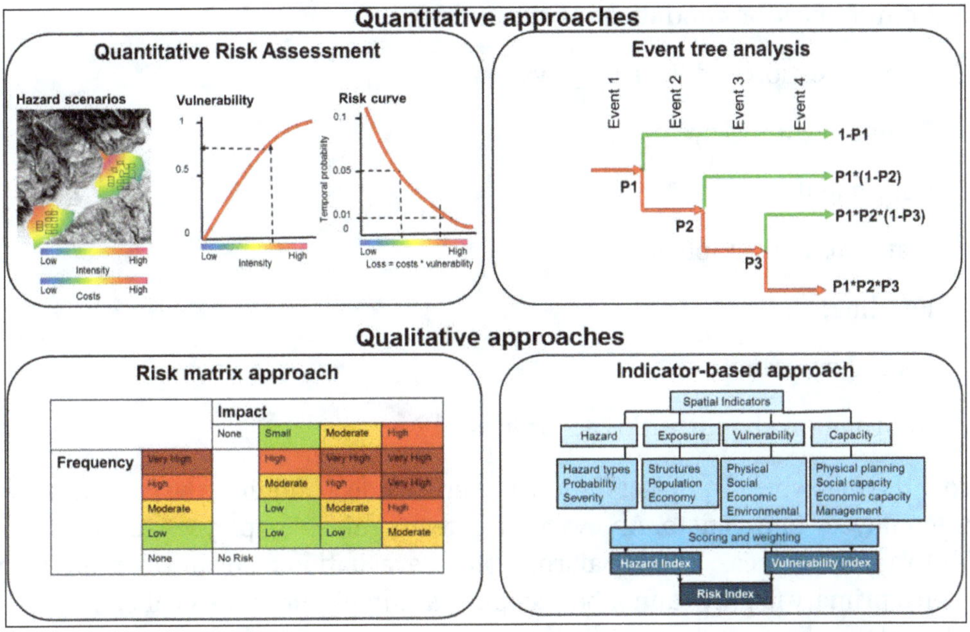

Components relevant for risk assessment, and the four major types of risk mapping.

Table: Advantages and disadvantages of the four risk assessment methods discussed:

Method	Advantages	Disadvantages
Quantitative risk assessment (QRA)	Provides quantitative risk information that can be used in Cost-benefit analysis of risk reduction measures.	Very data demanding. Difficult to quantify temporal probability, hazard intensity and vulnerability.
Event-tree analysis	Allow modelling of a sequence of events, and works well for domino effects.	The probabilities for the different nodes are difficult to assess, and spatial implementation is very difficult due to lack of data.
Risk matrix approach	Allows to express risk using classes instead of exact values, and is a good basis for discussing risk reduction measures.	The method doesn't give quantitative values that can be used in cost-benefit analysis of risk reduction measures. The assessment of impacts and frequencies is difficult, and one area might have different combinations of impacts and frequencies.
Indicator-based approach	Only method that allows to carry out a holistic risk assessment, including social, economic and environmental vulnerability and capacity.	The resulting risk is relative and doesn't provide information on actual expected losses.

Resource Control and Planning

There are many resources required for the preparedness program including:

- People,
- Facilities,

- Communications and warning technologies,
- Fire protection and life safety systems,
- Pollution control systems,
- Equipment,
- Materials and supplies,
- Funding,
- Special expertise,
- Information about the threats or hazards.

Consider the following examples: If there is a fire inside a building, the fire alarm system warns employees to evacuate. An evacuation team guides employees to safe exits and outside to assembly areas. The fire alarm system, evacuation team and exits are resources. When a primary facility cannot be occupied, a suitable alternate facility may be used. The alternate facility is a resource for the business continuity plan.

Needs Assessment

A needs assessment should be conducted to determine resources needed. Resources may come from within the business including trained employees, protection and safety systems, communications equipment and other facilities owned or leased by the business. Other resources from external sources include public emergency services, business partners, vendors and contractors. The availability and capability of resources must be determined - some are required immediately. For example, trained people (employees or public emergency services) capable of administering first aid or cardiopulmonary resuscitation (CPR) must be available to respond at a moment's notice. Other resources such as plywood to board up windows in anticipation of a hurricane may be stockpiled in advance or purchased when a storm is forecast. Even if plywood is stockpiled in advance, temporary labor may be needed to install the plywood over windows and doors. The availability of resources often depends on logistics. Logistics is the management of resources to get them to where they are needed when they are needed. Assessing resources for the preparedness program begins with reviewing program goals and performance objectives. High-level goals of the program include:

- Protect the safety of employees, visitors, contractors and others who may be at risk from hazards at the facility.
- Maintain customer service by minimizing disruptions of business operations.
- Protect facilities, physical assets and electronic information.
- Prevent environmental pollution.

- Protect the organization's brand, image, and reputation.

Examples of performance objectives include:

- The first aid team (that is trained to administer first aid and perform CPR) will be able to reach any employee within two minutes.
- The evacuation team will be able to direct all employees to safe exits and account for them outside the building within four minutes.
- Customer service staff will begin contacting customers within 8 hours of a service disruption using office space and telephone service provided by a business partner.
- The primary network server will be restored within 24 hours with replacement equipment from your primary vendor and data restored from backup media retrieved from the secure storage site.
- Production of product A will resume within 1 week by displacing production of product B at Plant B.

For each objective, an assessment of resources needed to accomplish the objective should be conducted. Simple objectives may require limited resources. Aggressive objectives will require many resources with significant capabilities available on short notice. Remember, without sufficient resources, or if resources lack required capabilities, objectives may not be attainable.

Conducting the needs Assessment

Besides identifying specific resources for the preparedness program, the needs assessment should answer other questions:

- What quantity of a resource is required?
- When will the resource be needed?
- What capability does the resource need to have? Are there any limitations?
- What is the cost for procuring or having the resource available? Are there any liabilities associated with use of the resource?

Resources

There are many resources needed to support the preparedness program. These resources can be organized into different categories:

- People.
- Facilities.

- Systems.
- Equipment.
- Materials.
- Supplies.
- Funding.
- Information.

Resources are needed for all phases of the program including prevention/deterrence, mitigation, emergency response, business continuity, crisis communications and disaster recovery.

Human Resources

Employees are needed to staff emergency response, business continuity and crisis communications teams. The emergency response team may be limited to employees trained to direct evacuation or sheltering. Some businesses may choose to organize emergency response teams to administer first aid, perform CPR and use automated external defibrillators (AEDs). Still others may train staff to use portable fire extinguishers. Regulations define minimum requirements that include training and organizing employees. Staff is needed to develop and manage the business continuity and crisis communication plans. The teams will likely be made up of employees working in their respective departments. Some staff may be assigned to work at alternate worksites if a primary worksite cannot be occupied.

Facilities

Facilities for emergency response include defined shelter space for protection from a tornado or interior space when "shelter-in-place" from an exterior airborne hazard is required. Facilities should also include a room that can be equipped to serve as an emergency operations center for supporting response to an incident. Other facilities needed include office space or a meeting room with communications equipment to serve as a communications hub. Facilities for business continuity may include alternate workspace equipped for continuation of business operations. Alternate facilities may be owned or contracted including office space, data center, manufacturing and distribution.

Systems

Systems for emergency response may include detection, alarm, warning, communications, suppression and pollution control systems. Protection of critical equipment within a data center may include sensors monitoring heat, humidity and attempts to penetrate computer firewalls. Every building has exit routes so people can evacuate if there is a hazard within the building. These exit routes should be designed and maintained in accordance with applicable regulations. Business continuity resources may include spare or redundant systems that serve as a backup in case primary systems fail. Systems for crisis communications may include existing voice and data technology for communicating with customers, employees and others.

Equipment

Equipment includes the means for teams to communicate. Radios, smartphones, wired telephone and pagers may be required to alert team members to respond, to notify public agencies or contractors and to communicate with other team members to manage an incident. Other equipment depends on the functions of the team. Automated External Defibrillators may be required for a first aid/CPR team. Fire extinguishers would be required for a fire brigade. Spill containment and absorbent equipment would be required for a hazardous materials response team or trained employees working in their assigned workspace. Personal protective equipment including hearing, eye, face and foot protection may be required for employees as part of a safety program. Many tools may be required to prepare a facility for a forecast event such as a hurricane, flooding or severe winter storm.

Materials and Supplies

Materials and supplies are needed to support members of emergency response, business continuity and crisis communications teams. Food and water are basic provisions. Systems and equipment needed to support the preparedness program require fuel. Emergency generators and diesel engine driven fire pumps should have a fuel supply that meets national standards or local regulatory requirements. That means not allowing the fuel supply to run low because replenishment may not be possible during an emergency. Spare batteries for portable radios and chargers for smartphones and other communications devices should be available.

Funding

Money invested in the preparedness program can pay big dividends if an incident occurs. Consider the benefits of a fire being controlled quickly; immediate medical assistance that saves an injured employee; or a recovery strategy that enables continued customer service. Spending funds prudently on preparedness can pay back multiple times when measured against the potential for damage to equipment, facilities, loss of staff, lost customers and lost revenue.

Internal Resources

There are many resources within your business that are needed for your preparedness program. These internal resources include staff for emergency response, business continuity and crisis communications teams. Other resources include facilities, systems, equipment, materials and supplies to support response, continuity and recovery operations. Identify needed resources and determine what resources are available internally. Resources that are not available must be obtained from external resources. Consider the following internal resources for your preparedness program.

People

Employees can be assigned the following tasks:

- Monitor weather forecasts and Emergency Alert System messages, broadcast warnings if severe weather is approaching or other warnings are broadcast, and alert the emergency response team.
- Direct evacuation and shelter actions.
- Administer first aid, CPR and use automated external defibrillators (AEDs).
- Provide facility security and take the lead on threats including bomb threats and suspicious packages.
- Operate building detection, alarm, communications, warning, protection and utility systems.
- Stabilize an incident using fire extinguishers; or cleaning up/containing small spills of hazardous chemicals.
- Prepare a facility for a forecast event such as severe weather.
- Clean up damage following an incident.
- Lead the business continuity team; provide support for the team.
- Execute recovery strategies for critical or time sensitive business processes.
- Serve as a spokesperson as part of the crisis communications team; communicate with employees, stakeholders and the news media; answer requests for information.

Employees should be trained so they understand the importance of their assignments and follow established procedures. Some employees may be given the opportunity to learn new skills.

Facilities

Office space and meeting rooms can be used as an emergency operations center (EOC), which is a facility for incident management. The EOC is a place to bring together personnel, gather information, facilitate communications, procure resources and support preparedness, response, continuity and recovery efforts. Rooms or areas within the interior of a building that are structurally strong can shelter people from a tornado. Unobstructed exits that are marked with signs and equipped with emergency lighting are essential to quickly evacuating people if there is a fire or hazard inside. Owned buildings at another site may be used as alternate workspace if a building cannot be occupied. This depends upon the location of the building and whether the building would be affected by the same hazard that prevented use of the primary building. The alternate facility may be a viable business recovery strategy if the building can be configured with the required equipment or existing equipment can be configured to need business requirements.

Systems and Equipment

Many systems and equipment are needed to detect potential hazards and threats, protect life safety and property and continue business operations. These resources include:

- Detection systems (fire detection, burglar alarm or intrusion detection, computer network security, Emergency Alert System receivers and television, radio, for news and weather).
- Alarm systems (fire alarm, intrusion alarm and process system alarms).
- Warning systems (occupant warning systems include fire alarm, public address and tornado warning).
- Communications systems (landline telephones, cellphones, smartphones, email and data, radios and pagers).
- Pollution containment systems (primary and secondary building containment and devices to stop the flow of materials from tanks and piping).
- Fire protection and suppression systems (fire sprinklers, fire extinguishers, fire pumps and water supplies, special extinguishers for computer rooms and special hazards).
- Emergency power supplies (uninterruptible power supplies and generators).
- Building utility systems (electrical, plumbing, heating, ventilation, air conditioning and sanitary).

Evaluate these systems to determine whether they meet the needs of the program. Identify and plan to overcome emergency communication system limitations such as weak radio or cellular service or areas where a warning system cannot be heard. Upgrading this critically important system may be required. Verify that these systems are in reliable working condition. If fuel, battery backup power or batteries are required, make sure the system can run for the required time and chargers are available. Document how to operate these systems and mark the locations of controls. Make sure the information is available during an emergency. Many of these systems also require periodic inspection, testing and maintenance in accordance with national codes and standards. Train staff so a knowledgeable person is able to operate systems and equipment.

Materials and Supplies

There are lots of basic materials and supplies needed for the preparedness program. These "consumable" resources include clipboards, paper forms, pens and pencils. Sufficient copies of paper forms are especially important to do automated tasks manually. Flashlights with spare batteries are needed if the power goes out. Provision of food and water for personnel engaged in preparedness, response, continuity and recovery activities should also be addressed in the plan.

External Resources

Preparing for an emergency, responding to an emergency, executing business recovery strategies and other activities require resources that come from outside the business. If there were a fire in the building, you would call the fire department. Contractors and vendors may be needed to prepare a facility for a forecast storm or to help repair and restore a building, systems or equipment following an incident. An understanding of the availability and capabilities of external resources is needed to make decisions about the preparedness program. How long would it take the fire department to arrive? How do you reach a contractor late at night and how long will it take them to arrive? Determination of the response time and capabilities of external resources will help you identify gaps between what you need and what is available. Strategies should be developed to fill these gaps. The following external resources should be identified within plan documents. Include contact information to reach them during an emergency and any additional instructions within the preparedness plan.

Public Emergency Services

- Fire,
- Emergency medical services,
- Hospital or emergency health care provider,
- Rescue,
- Hazardous materials,
- Law enforcement (local, county, state police),
- Public health,
- Public works.

Contractors and Vendors

- Emergency services (hazardous materials cleanup, facility repair and restoration).
- Systems and equipment (procurement, inspection, testing and maintenance).
- Information technology (equipment procurement, data backup, recovery solutions).
- Business continuity (generators, temporary equipment, leased space, office trailers).

Partnerships

- Business partners (suppliers, contractors, vendors and professional services firms that could lend assistance with services, temporary workspace and other resources).
- Businesses or civic organizations in the community.

Logistics

Logistics considerations are an important part of the preparedness program to ensure that resources will be available when and where they are needed. Compile an inventory of internal and external resources to identify their location, the operating procedures and the persons who can operate these systems. Also, note the estimated delivery or response time of external resources. A person should be assigned responsibility for logistics and to manage resources to support the preparedness program. They should work with the emergency response and business continuity teams who can identify resource needs. Logistics procedures should define procurement requirements including the names of employees who have the authority to issue purchase orders and contract for services. Procedures should also be established to expedite obtaining resources during an emergency. Open purchase orders with potential contractors and vendors will expedite the procurement process.

Thunderstorm and Tornado Preparedness

Though thunderstorms are dangerous, their power is often underestimated because they're so common. Everyone experiences thunderstorms regularly, and that makes it easy to forget their unpredictability. Violent storms can cause a variety of other dangerous events, here are a few:

- Lightning,
- Strong winds,
- Microbursts,

- Flash floods,
- Hail,
- Tornadoes,
- Wildfires due to lightning strikes.

The number-one thing you can do to protect you, your family, and your belongings from the dangers of thunderstorms – or any natural disaster for that reason – is to be prepared.

Being prepared means having the knowledge and supplies necessary to stay safe and handle emergency situations to the best of your ability. It doesn't always mean you're going to be safe, but it does improve your chances.

Building an Emergency Preparedness Kit

The first thing you need to do to be prepared is have an emergency preparedness kit. This kit contains everything you and your family will need in an emergency, as well as enough supplies to survive for three to five days.

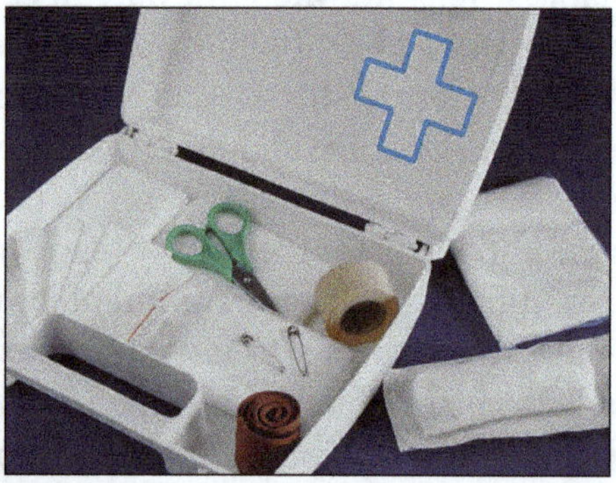

Here's a list of items that every emergency preparedness kit should contain:

- One gallon of water per person, per day.
- Around 2,000 calories of non-perishable food per person, per day.
- A mechanical can opener.
- First-aid supplies to handle minor injuries, including splints, suture kit, anti-bacterial creams, bandages, and antihistamine.
- Hand-crank or battery-operated radio.
- A week's worth of all prescription medications.
- Hygiene items such as toothpaste, shampoo, and feminine products.

- Copies of important documents in a waterproof container; be sure to include driver's licenses, birth certificates, social security cards, and insurance cards.
- Blankets.
- Flashlights and extra batteries.
- A minimal amount of money in small bills.
- An extra credit card.
- Emergency contact list, including names, telephone numbers, and addresses.
- If you have children, include items to entertain them, such as coloring books, cards, or board games.
- If you have an infant, be sure to include diapers, formula, and anything else you'll need.
- If you have pets, have enough food and other supplies for them.

For the elderly or anyone with a disability, make sure to include items to meet their specific needs.

Preparedness before Severe Weather Hits

Preparation for severe weather, including thunderstorms and tornadoes, happens well before a storm is spotted. To be ready and stay safe, here are things you should do on a regular basis, before a storm hits:

Landscaping

Keeping your yard clean of dead trees and rotting limbs protects both you and your property. It reduces the risk of limbs falling onto a person, vehicle, or even your home.

If a tree is starting to die, removing it from the property guarantees it won't uproot or snap and fall during high winds.

Secure Outdoor Objects

When there's a chance of severe weather, secure your outdoor items to the ground. From picnic tables to fuel tanks, high winds can topple and blow huge items around, causing damage and making cleanup more difficult. Take small, easy-to-move items indoors.

Watch vs. Warning

If your area is at risk of severe weather, know what to be prepared for. The weather radios keep you informed of National Weather Service weather alerts. A severe weather watch means that conditions are right for severe weather to occur. A severe weather warning means that a thunderstorm has been reported or indicated by radar.

Just to clarify:

- Watch = Possibility.
- Warning = Occurring.

Keep Activities Inside

If there's a thunderstorm risk, postpone any outdoor activities or make alternate plans. You don't need 15 kids outside during a thunderstorm for a birthday party. Either have it another day or have a plan B just in case.

Protect Pets and Livestock

While it may seem that pets and livestock should be able to withstand a thunderstorm, animals should be put indoors and out of harm's way. They are susceptible to being struck by lightning, especially those who reside in open fields. Hail can also be dangerous to livestock without cover, and can cause injuries or even be fatal. To keep them safe, secure livestock in a sturdy building, with plenty of food and water.

Preparedness during a Thunderstorm

The most severe storms occur when one storm lingers for an extended time over a specific area. Only 10 percent of all storms are considered severe, which indicates one of the following:

- Hail with a diameter of at least one inch.
- 58-mph winds or greater.
- Produces a tornado.

Because of their unpredictability, once a thunderstorm is in the area, there's not much that can be done to prepare for the damage that may occur. Even so, there are things you can do to increase your chance of safety.

Head Indoors

If you can hear thunder, you're within reach of lightning – which can travel up to 10 miles outside of the storm. Hence, the lack of rain or thunder is in the distance is not a sign of safety. As soon as a storm is in your area, it's time to head inside. And inside means completely indoors, not on the back porch and not in the front door. While thunderstorms can be beautiful to watch, they're dangerous. Even standing in front of a window can lead to unnecessary injuries.

Regardless of what urban myth you've heard, rubber soles and tires do not protect you from lightning, so don't think wearing sneakers means it's okay to continue your golf game. It's not – ever.

Shutter Windows

Shutter your windows when the storm starts in order to protect them from high winds, hail, and debris. If you don't have shutters, close the drapes or blinds for added protection. Be sure to stand away from other glass structures, including glass doors and skylights.

Unplug Electronics

Unplug electronic devices – including computers, television sets, and other items – that may be damaged if there's a power surge or lightning strike. Avoid using items that plug in, as current can travel through them and shock you.

Avoid Corded Phones

If you need to make a phone call during a thunderstorm, do NOT use a corded phone or a cell phone that is plugged in. Electricity easily travels through phone lines and can actually electrocute you through the phone.

Avoid Plumbing

A thunderstorm is not the time to take a shower or do the dishes. The shock from lightning can travel through the water and pipes, making faucets and fixtures dangerous.

Avoid Concrete

When it's storming, be sure to avoid laying on concrete floors or leaning on concrete walls. While concrete itself does not conduct electricity, it's often reinforced with steel, which does.

Be Aware of Natural Lightning Rods

Lightning often strikes the highest thing available, so be aware of solitary trees, TV antennas, power lines, or other high structures. If outside, try to make yourself smaller than surrounding objects to avoid being struck by lightning.

If in a Vehicle

If you're in a vehicle when a storm hits, turn on your flashers and pull over until you can safely see to drive. While the rubber tires of the vehicle do not protect you from lightning, the hard metal top does and makes your vehicle the second safest place to be during a thunderstorm. Just be sure you're not touching any metal in the vehicle, and if you're in a soft-top vehicle, seek shelter as soon as possible.

If in the Woods

If you're in the woods when severe weather hits and are too far away to get to a building or vehicle, seek shelter in a low area, preferably under a thick cover of small trees. Try to avoid dead trees or hanging limbs.

If in the Open

If you're in the open with no shelters or vehicles nearby, get as low as you can in a ravine or valley. Crouch down, staying on the balls of your feet, and cover your head, making yourself the smallest target possible. While this protects you from lightning and winds, be cautious of flash floods, which cause the most storm-related deaths in the U.S.

If in Open Water

Being in open water during a thunderstorm is perhaps one of the most dangerous places to be. Head toward land as soon as signs of a storm appear. Stay away from metal on the boat and phone in for assistance if necessary.

The Risks of Lightning

Lightning is one of the main reasons that thunderstorms are so dangerous, and is one of the top three causes for storm-related fatalities in the U.S., with an average of 51 deaths per year and hundreds of injuries.

Here are some lightning facts that are important to know:

- Lightning occurs when there is a buildup and discharge of electrical energy.
- While anyone can get struck by lightning, over 80 percent of fatalities are men, aged 15 to 40.

- The chance of getting struck by lightning is one in 600,000.
- Most lightning deaths and injuries occur in the summer when someone is stuck outside during a storm in the afternoon or evening.
- Lightning heats the air around it to 50,000 degrees Fahrenheit, which is hotter than the surface of the sun.
- Because this air is so hot, it produces a shockwave, which we hear as thunder.
- There is no such thing as heat lightning; it's actually regular lightning from a thunderstorm that's too far away to be seen.
- It doesn't need to be raining for there to be lightning; dry lightning is particularly common in the Southwest, where the rain evaporates before hitting the ground.
- To figure out how far lightning is, count the time between seeing the strike and hearing the thunder; every five seconds is about one mile.

There is a risk of being struck by lightning when any of the following occur:

- You see lightning.
- You hear thunder.
- There's a loud static on AM radio.
- There's a buzz sound on the radio antenna.
- When the masthead on a ship begins to glow, which is also called St. Elmo's Fire.

If Struck by Lightning

If you're with someone who's been struck by lightning, here's what you need to do to keep them safe:

- A person struck by lightning does not hold an electric charge, so don't be afraid to touch them.
- Call 911 immediately.
- Look to see if the individual is breathing; if not, perform mouth to mouth.
- Check for a heartbeat; if you can't find it, start CPR.
- Look for burns where the lightning entered and exited the body.
- There may also be nervous system damage, broken bones, or trouble with hearing and eyesight.

Tornado Preparedness Checklist

One of the most dangerous aspects of thunderstorms is their ability to spawn tornadoes. A rotating column of air extending from a cumuliform cloud, tornadoes are destructive forces that can be over a mile wide and extend for over 50 miles with winds in excess of 200 mph. While tornadoes can happen anywhere in the world, they most often occur in the U.S., with an average of about 1,200 per year.

Tornadoes typically occur when it's warm, humid, and windy, and can also be formed from hurricanes and tropical storms. While the tornado itself is extremely dangerous, most damage occurs from the strong winds and flying debris carried by the tornado.

Since tornadoes form from severe thunderstorms, the precautions associated with storms should also be followed for tornadoes. Tornado safety plans should always include the following:

Basement

Anytime there's a chance of a tornado, you want to do more than just head indoors – you want to get to the lowest floor possible, preferably underground. Once there, get under a workbench to protect yourself from falling debris. If that's not available, cover yourself with a mattress.

Interior Room

If your home does not have a basement, move to a small, interior room with no windows. Do not worry about opening up windows or leaving doors open. Most buildings have enough leaks that the air does not pressurize to the point of breaking glass.

If you don't have an appropriate interior room, hallways, under stairwells, and even bathtubs offer some protection during a tornado.

Safe Room

Often used in mobile home parks and community buildings, safe rooms are specifically designed to provide more safety during severe weather and tornadoes. These rooms are reinforced and designed to withstand high winds. If you do build a safe room, be sure to register it with your local fire department. That way if an emergency occurs and you get trapped in the room, the authorities will know where to look.

Assume the Position

Whether you're in a basement or school hallway, get into a crouched position against a wall or solid structure. Cover your head and neck with your arms, and if it's available, cover yourself with a mattress, blanket, or even towels.

Be sure to position yourself in a place that isn't below heavy objects on the above floors. Things like pianos and waterbeds have been known to fall through floors during tornadoes, and can cause significant and even fatal damage.

Mobile Homes

Mobile homes and travel trailers are NEVER safe places to be during a tornado. Even if the building is strapped to the ground, these structures are often lifted and thrown. If you live in a mobile home, be sure to have an outdoor safe room or a place to go underground in the chance of a tornado.

Vehicle

If you find yourself in a vehicle and a tornado is approaching, seek shelter in a sturdy building immediately. If that's not possible, you have a few different options depending on your situation:

- Outdrive it: If the tornado is far enough away and you think you can outrun it in your vehicle, do so. Move away from the tornado at 90 degree angles. Remember, tornadoes change directions frequently and without warning. It is NEVER safe to move toward a tornado for any reason.
- Secure yourself inside: If there's no building nearby and the tornado is too close to drive away, secure yourself inside the vehicle. Put your seatbelt on, and lower yourself so your head and body are below the windshield. Protect your head and neck by covering them with your arms, and cover yourself with a blanket or jacket if possible. Vehicles often get lifted and tossed by tornadoes, but securing yourself inside lowers your risk of injury or death.
- Get lower: The only time you should leave your vehicle during a tornado is if you can get lower than the roadway. If there's a ditch or ravine, get in it, and cover your head. If there's something secured to the ground that you can get near or hold on to, do so. Be sure to avoid areas where things could topple on top of you. Bridges and underpasses are NOT safe and too often collapse to offer any safety.

Outdoors

If you're stuck outside during a tornado and there's no building or vehicle to get to, get in the lowest place you can find. A ditch, valley, or ravine offer some protection against the elements. Try to avoid places with debris that can be picked up by the tornado.

Pets and Livestock

During severe weather, animal behaviors change. They become nervous, agitated, and may be impossible to keep calm. Animals often sense tornadoes well before people do, so be prepared that it may be more difficult than normal to handle them.

If a tornado is in the immediate vicinity, do not worry about your livestock. Animals are not worth risking your own or your family's safety. That being said, if time allows, here are some things you can do to increase safety for your livestock during a tornado:

- Never leave animals tied up or restrained outside.
- If time is limited, open up escape routes, allowing your animals to get to safety on their own.
- If you have a barn or building that is secure, bring animals into the shelter and provide food and water before the storm hits.

Preparedness after a Tornado

After a tornado passes, it's time to evaluate the damage and offer assistance to those who need it. First and foremost, seek medical attention for those who are injured. Emergency personnel may have trouble reaching people, so keep your emergency preparedness kit with you.

If your home is still standing, don't assume everything is okay. Buildings get hit with projectiles and often suffer structural weaknesses during tornadoes. Just because it's still standing does not mean that it's safe. Proceed with caution.

Tornadoes can also expose homes, water supplies, and communities to hazardous materials. Be sure to talk to local authorities about proceeding when this occurs.

Food Safety in Emergency Situations

In emergency situations, difficult conditions can be made even worse in the event of foodborne disease outbreak. Provision of food is often carried out under crude conditions and the people available to organize and implement feeding operations may not have the necessary experience to manage food safety effectively.

Preliminary Issues

Planning

Food safety needs to be considered at the planning stage of any mass feeding or food distribution activity. In emergency situations, the following factors may have particular significance.

- Scale and demographics of operation:
    - Poor health and nutritional status may lead to increased susceptibility to foodborne illness.
    - Very young and very old people may also be more vulnerable to food related illness.
    - Preparing and handling food for large numbers of people under difficult conditions involves logistical difficulties and calls for a high degree of planning to minimise risks, particularly microbiological risks.
    - Cultural acceptability of methods of preservation or preparation (under emergency situations, people may be traumatised and more sensitive to unfamiliar products).
- Conditions on the ground: The availability of power/fuel, clean water, storage facilities, sanitation, etc. will impact the type of food that can be distributed safely. Shelf-life and the need for refrigeration or other special storage conditions of foods should be a consideration in planning the operation.
- Assembling the team: Include a team member with some food safety training who can provide rapid on the ground training/briefing to food handlers and has sufficient knowledge on food hazards, regulations and standards to accept general responsibility for food safety issues and provide advice to the operations manager on food safety concerns, issues, and necessary actions.

Interactions with other Functions Involved in Emergency Situations

Mass feeding or food distribution activities are part of a larger emergency response team: the activities of other groups have important implications for effective food safety management. The person responsible for overseeing food safety issues in the mass feeding exercise should interact with groups carrying out other functions so as to ensure coordination and information flow as required. The following 'functions' have some relevance to food safety.

- Medical: To facilitate reporting of illnesses, health checks of food handlers.
- Nutritionist: To ensure that safe foods are also adequate from the nutritional perspective.
- Transportation: To ensure conditions for hygienic transport of foods.
- Power supply: To be informed on availability and disruptions in supply or to inform of specific power requirements
- Water management: To ensure and adequate supply of potable water for feeding stations; and adequate facilities for its hygienic storage and distribution.
- Pest control: To protect food supplies from damage and contamination from pests; and to minimise risk of chemical contamination from pesticides being used in the vicinity of food, feeding operations, water supplies, etc.
- Waste management and sanitation: To minimize risks of contamination from the environment – directly or through the harbouring of insects and other pests that can be vectors for the transmission of contamination. Relevant activities include garbage and waste disposal, control of entry or movement of animals that might pose hygiene problems and ensuring that temporary toilet facilities will not be a source of contamination of foods.
- Construction: To facilitate appropriate selection of location and construction of temporary facilities being established to house food-related operations.
- Security: To be informed on anticipated changes in the numbers of people to be fed; to ensure that there is restricted access to hygiene sensitive areas and to be informed about new emergency events and routes for evacuation of people if needed.

On-site Activities

Receiving and Storing Food

- Imported food/food aid should be checked to make sure that it meets agreed standards by:
    - Checking accompanying documentation against national requirements,

   - Inspection,
   - Checking conditions in which food was transported.
- Locally supplied food should be checked by:
   - Checking any accompanying documentation,
   - Inspection,
   - Checking conditions in which food was transported,
   - Requesting information from supplier, delivery people (particularly if documentation is unavailable) with regard to transport, storage, production, source, any local conditions or warnings that may indicate the potential for contamination of food/food sources etc.
- Some basic documentation and record keeping (e.g. on inspection outcome, accompanying documentation) is advised. (It may be possible to combine this with any other records already being kept at food distribution/feeding stations).

Instructions for Handling of Rejected Supplies

Supplies should be clearly labeled as rejected, separated from all other food supplies and disposed of or destroyed. Records of food rejections should be kept (what food, batch identification information, reason for rejection, date, person deciding on rejection, etc) and these should be reported to the appropriate authorities. Decisions regarding possible alternative use of rejected foods should be made in consultation with a food safety specialist.

Instructions for Handling of Suspect Supplies

Supplies should be clearly labeled as suspect and separated from all other food supplies. The decision on whether or not to use such supplies should be taken according to the need for the food, the nature of the problem with the food and the facilities available to prepare the food (e.g. decontamination by cooking, prevent cross-contamination of other foods). Such a decision should be made in consultation with a food safety specialist. Note that national analytical services are likely to be overwhelmed in the case of an emergency and therefore it is unlikely that decisions can be based on the results of laboratory analysis.

Instructions for Storage of Accepted Supplies

- The storage area should be dry with good ventilation (food aid may include dry grains, e.g. rice, maize lentils and dry storage is critical to prevent moisture, mould growth etc.).
- As far as possible, the structure should offer protection from entry of pests.

- Food supplies should not be in direct contact with the ground or the floor or walls. Pallets, branches, boards, bricks, concrete blocks or plastic can be used to keep supplies off the ground.

- Foods should be stored in a cool dark place but at least out of direct sunlight (foods such as oils and vitamin fortified products are highly susceptible to deterioration under unsuitable conditions of storage e.g. in sunlight).

- Food supplies should be stored in a separate area to chemical supplies (e.g. detergents, sanitizers, pest control agents).

- If possible, pest resistant containers should be used for food storage or some form of pest control/management implemented.

- Spilled food should be cleaned up immediately to discourage rodents and other pests.

- Raw foods should be stored separately from ready-to-eat foods. Furthermore, perishable foods such as fish and meat should be stored separately (either physically or by time with effective intermediary cleaning and disinfection) from other raw foodstuffs such as fruit and vegetables to prevent cross- contamination.

- If possible, perishable foods should be refrigerated. If refrigeration is not available, these foods need to be kept in as cool a place as possible out of direct sunlight and used quickly.

- Periodic checks of the storage conditions should be undertaken to ensure they are maintained. Foods, particularly those being stored for longer periods e.g. grain should also be inspected periodically for deterioration, infestation etc.

Basic Facility

While recognising that food distribution/mass feeding activities are carried out under crude conditions, the basic facility should provide for:

- Appropriate storage of food: Storage area must be protected from moisture and pests and be separate from storage of chemicals and other potential hazards.

- Location of the food preparation area and the storage area away from toilets, waste disposal areas, and other potential sources of contamination.

- Facilities for food preparation that minimise risk of contamination including cleanable food contact surfaces, floors, walls.

- Adequate supplies of clean and potable water.

- Supplies of power/fuel to enable cooking at required temperatures and times and to facilitate related operations (cleaning, storage).

Personnel

It is unlikely that many of the food handlers involved in emergency food handling operations will have received instruction of food safety. It is therefore advised that food handlers:

- Should receive instruction/guidance in basic hygiene (personal hygiene, food hygiene and safe food preparation and handling) according to their function, including information on the important role they play in the health of the people they are feeding.
- Should show no signs of illness (symptoms to watch out for include diarrhoea, vomiting, fever, jaundice, sore throat, discharge from the nose, ears or eyes, visibly infected skin lesions (cuts, boils, etc.).
- Should, if possible, have their health status regularly checked by medical staff to minimise transmission of disease.

In addition,

- The operation manager or a designated person should oversee food handling operations to identify any potential hazards and provide in-service orientation to food handlers while undertaking their activities to reinforce the food safety messages.

Hygienic Practice in Daily Operations

Water for Food Preparation

Only potable water should be used in food preparation. This can be achieved through boiling and disinfection.

Temperature-time Issues

Cooking food: Food should be cooked thoroughly to kill pathogenic bacteria. Most are killed above 70 °C but some can survive to 100 °C and above. (Some bacteria produce heat resistant spores which survive and possibly germinate at high temperatures. It is therefore important to keep hot foods hot (above 60-65 °C) to prevent growth of such bacteria.

The following indicators can be used to ensure food reaches a sufficiently high temperature and is well cooked:

- The food is steaming hot throughout.

- Meat and poultry juices run clear and the meat itself contains no pink or red parts.
- Soups, stews and other liquid based foods are held at a rolling boil for 15 minutes.

Under emergency or difficult conditions and high ambient temperatures foods can have a higher level of microorganisms and so should be cooked for longer than normal to minimize the risk while also maintaining acceptable organoleptic qualities for consumption. Perishable foods that have been left at ambient temperature for prolonged periods may contain pre-formed, heat stable toxins that are not eliminated by cooking.

Holding food: The danger temperature zone for microbial growth in food is 5 – 65 °C. Before preparation/cooking:

- Holding of food within the danger zone prior to cooking will lead to increased microbial loads and possible toxin production. While increased cooking times will adequately reduce microbial loads, it will not resolve the problem of toxin accumulation.
- Minimize the time at which perishable foods are held at ambient temperatures before cooking.

After preparation:

- Food should be eaten immediately after preparation and in no case should cooked food be left at ambient temperatures for more than 2 hours before consumption.
- Hot food should be kept above 60-65 °C (steaming) before serving. It should be covered to help retain heat.
- If there is no refrigeration or food storage available, food should be cooked as needed to prevent leftovers (This is an important link to the logistical management of the feeding station and information on the quantities of food to be prepared and associated scheduling).

Foods requiring refrigeration/perishable foods: In most difficult conditions it is probably better to minimise the use of foods requiring refrigeration as breaks in the 'cold chain' are probably inevitable. If such foods are used, then the following precautions are recommended:

- Use as quickly as possible.
- Store in as cool a place as possible and away from direct light (Remember the danger temperature zone for microbial growth is 5 – 65 °C).

Frozen foods: Facilities for frozen storage are not likely to be available in most emergency situations, and even if there are, interruptions to the 'cold chain' are very likely. In situations where frozen food is being use, the following precautions are recommended:

- Defrost at a cool temperature to minimise bacterial growth on the surface.
- Ensure that frozen block sizes facilitate defrosting within required time.
- Where the above is not feasible frozen food which is available in small pieces can be cooked directly without defrosting.
- Defrosting can be accelerated through the immersion of packaged frozen food in water with agitation (only if potable water is available).

In cases where the cold chain has been interrupted, the integrity of the frozen food should be assessed. In general the food is still safe to use if:

- It still contains ice crystals.
- Feels as cold as if refrigerated.

Such food should be cooked or re-frozen immediately.

Avoiding Food Contamination

Food can be contaminated from the environment, personnel, food preparation surfaces and utensils, raw and uncooked food, animals, pests and waste material. To prevent contamination, control over the various operations related to food preparation need to be exercised.

Prevention of microbiological contamination is an important function in food preparation and the focus of the guidance below. However, attention also needs to be given to possible physical and chemical contamination. Physical contaminants include stones, pieces of glass, metal and may be identified in the physical inspection on receipt of food. However, if food preparation is taking place outdoors or in temporary facilities ongoing vigilance is important to prevent physical contamination of food. All chemicals (detergent, disinfectant, sanitizer) used in the food preparation area should be removed before food preparation begins to prevent any chemical contamination of the food.

In Relation to Personal Hygiene

- Thorough hand washing before and during food preparation especially after using the toilet, handling raw food or waste is critical.
- Soap, sanitizer and clean water should be available for hand washing at convenient locations.
- Ensure food preparation area is limited to dedicated food handlers (whose health status has been checked).

In relation to the food preparation area including food contact surfaces, utensils etc:

- There should be separate areas for raw and cooked foods (raw foods are a source of microorganisms which can recontaminate prepared foods).
- The food preparation area should be cleaned using hot water, where possible, and detergent to remove visible soil/dirt/food particles/grease, rinsed and then sanitized. Diluted bleach (hypochlorite solution) can be used to sanitize surfaces.
- Utensils and dishes etc. should be cleaned using hot water, where possible, and detergent to remove visible soil/dirt/food particles/grease, rinsed and then sanitized using boiling water or diluted bleach. (Bleach solutions should be regularly replenished to ensure effective sanitization).
- Cloths/sponges can harbour bacteria and require regular sanitizing by boiling, or using bleach.

Food particles/dirt that remain in the food preparation area present an important hygiene risk so effective cleaning routines are essential to minimize contamination.

Preparation of Bleach Solutions for Sanitization

For non porous surfaces: 5ml/litre.

For porous surfaces: 15 ml/litre.

For cloths/sponges: 60ml/litre (disinfecting solution). This can also be used to disinfect floors and waste containers. Follow with rinsing and allow to air dry.

Surfaces should be exposed to the bleach solution and utensils should be soaked in the solution for 2 minutes after cleaning. Non porous surfaces (metal, ceramic, glass hard plastic) and equipment should then be allowed to dry. Porous surfaces and utensils (wood, rubber, soft plastic), should be rinsed with clean potable water after sanitization and then allowed to air dry. Cloths/sponges should then be rinsed and allowed to air dry.

In relation to waste management:

- Personnel should be assigned to ensure that waste is regularly removed from the food preparation area.
- A means of disposing of both solid and liquid waste should be available for the food preparation.
- The waste disposal area should be located away from the food preparation area.
- Attention should be given to pest control in the waste storage area.

- Waste containers in the food preparation area should be cleaned regularly with water and detergent and disinfected with bleach solution.
- Every effort to be made to manage food preparation to keep waste food to a minimum to reduce the additional burden of waste disposal.

If waste is not adequately managed, including waste from toilets/latrines set in the vicinity of the food preparation area, there can be elevated levels of contamination and contaminant vectors in the environment that greatly increase the opportunity for contaminated food.

Food Safety, Nutrition and Vulnerable Population Groups

Actions to enhance food safety must also take into account the nutritional needs and possible vulnerable condition of the consumers.

When foods require rehydration, potable water should be used to prepare and rehydrate foods, and it is important to use the correct ratio of potable water to food to ensure correct nutrient balance.

When using water purified by boiling to prepare infant formula or other fortified dried foods it is important to allow the water to cool back to about 50 °C to minimise nutrient loss due to boiling water.

Food Supply of General Population

In the initial stages of an emergency, decisions will need to be made to assess the suitability and safety of existing food supply, determine appropriate means to ensure foods in the final stages of harvesting can be preserved as a future supply of foods, and whether of not certain foods are unfit for human consumption. This will require an assessment of the risks at the primary production level through processing and retail and storage facilities.

Assessment of Risks at the Primary Production Level

An assessment of the primary production is important from a food safety perspective given its importance as a source of food contamination. Flooding will increase the contamination levels through environmental contamination, destruction of infrastructure, contamination of feed and water supplies, contamination of aquaculture ponds and enhancing conditions for moisture accumulation, decay/spoilage and spread of disease.

Assessment of Retail and Storage Facilities

An assessment of food retail and storage facilities can provide an indication of the likely condition of foods that have been held there as well as possibly identifying structures that can be used for emergency food stocks.

Assessment of Food Handling/Processing

Operation of existing food handling/processing facilities may play an important role in ensuring the continuity of the food supply to the general population. Assessment of these facilities is necessary to evaluate whether they can be used or what repairs are required to make them operational.

Survival Kit

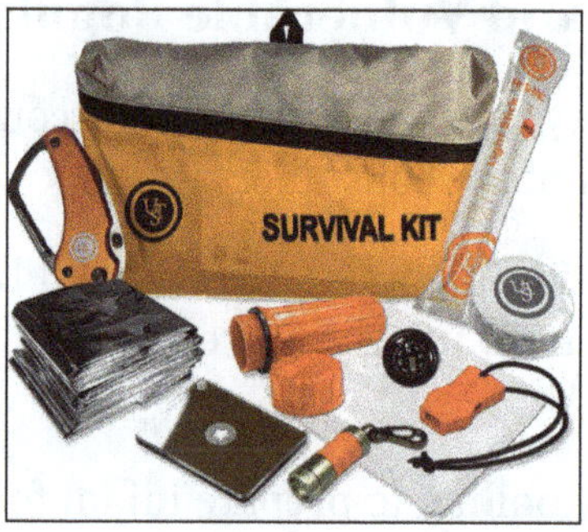

A survival kit is a package of basic tools and supplies prepared in advance as an aid to survival in an emergency. Military aircraft, lifeboats, and spacecraft are equipped with survival kits. Survival kits, in a variety of sizes, contain supplies and tools to provide a survivor with basic shelter against the elements, help him or her to keep warm, meet basic health and first aid needs, provide food and water, signal to rescuers, and assist in finding the way back to help. Supplies in a survival kit normally contain a knife, matches, tinder, first aid kit, bandana, fish hooks, sewing kit, and a flashlight. Civilians such as forestry workers, surveyors, or bush pilots, who work in remote locations or in regions with extreme climate conditions may also be equipped with survival kits. Disaster supplies are also kept on hand by those who live in areas prone to earthquakes or other natural disasters. For the average citizen to practice disaster preparedness, some towns will have survival stores to keep survival supplies in stock. The American Red Cross recommends an emergency preparedness kit that is easy to carry and use in the event of an emergency or disaster.

Humanitarian Aid

Humanitarian aid means predominantly life-saving and protection during disasters and crises, both natural and man-made. Humanitarian aid consists in:

- Providing assistance and support to people exposed to protracted crises as well as forgotten crises,

- Carrying out post-disaster and post-crisis rehabilitation projects, in cooperation with local institutions, taking into consideration long-term development goals,
- Overcoming difficulties associated with migrations caused by natural disasters or crises caused by human activity,
- Disasters prevention and mitigation of their negative effects.

Funds for humanitarian aid are sourced from the state budget target reserve administered by the Development Co-operation Department of the Ministry of Foreign Affairs. In the case of the most severe crises, other state institutions may also make a decision to grant humanitarian aid and fund it from their budgetary resources in coordination with the MFA.

Objectives of Humanitarian Action

The principal objective of humanitarian aid is to save and sustain life as well as to maintain human dignity of people in danger and those affected by both natural and man-made disasters. Humanitarian aid also encompasses actions aimed at disaster prevention, strengthening disaster preparedness and disaster risk reduction.

References

- What-is-emergency-management: uasem.org, Retrieved 26 January, 2020
- Prepare-disaster-preparedness-plan, environment -1744: preservearticles.com, Retrieved 05 March, 2020
- Methodology-51: charim.net, Retrieved 19 April, 2020
- Thunderstorm-tornado-preparedness-guide: ammo.com, Retrieved 25 May, 2020
- Survival-kit: educalingo.com, Retrieved 18 January, 2020
- Humanitarian-aid-251: polskapomoc.gov.pl, Retrieved 05 August, 2020
- Write-a-brief-note-on-disaster-preparedness-for-floods-2353: preservearticles.com, Retrieved 28 April, 2020
- Write-a-brief-note-on-cyclone-preparedness-2356: preservearticles.com, Retrieved 08 July, 2020

2

Disasters and their Management

There are various geophysical, hydrological, climatological, meteorological and biological disasters that fall under emergency management. It includes earthquakes, landslides, folds, drought, cyclones, disease epidemics, etc. The topics elaborated in this chapter will help in gaining a better perspective of these various types of disasters and their management.

Earthquake

Earthquake is any sudden shaking of the ground caused by the passage of seismic waves through Earth's rocks. Seismic waves are produced when some form of energy stored in Earth's crust is suddenly released, usually when masses of rock straining against one another suddenly fracture and "slip." Earthquakes occur most often along geologic faults, narrow zones where rock masses move in relation to one another. The major fault lines of the world are located at the fringes of the huge tectonic plates that make up Earth's crust.

Building knocked off its foundation by the January 1995 earthquake in Kobe, Japan.

Little was understood about earthquakes until the emergence of seismology at the beginning of the 20th century. Seismology, which involves the scientific study of all aspects of earthquakes, has yielded answers to such long-standing questions as why and how earthquakes occur.

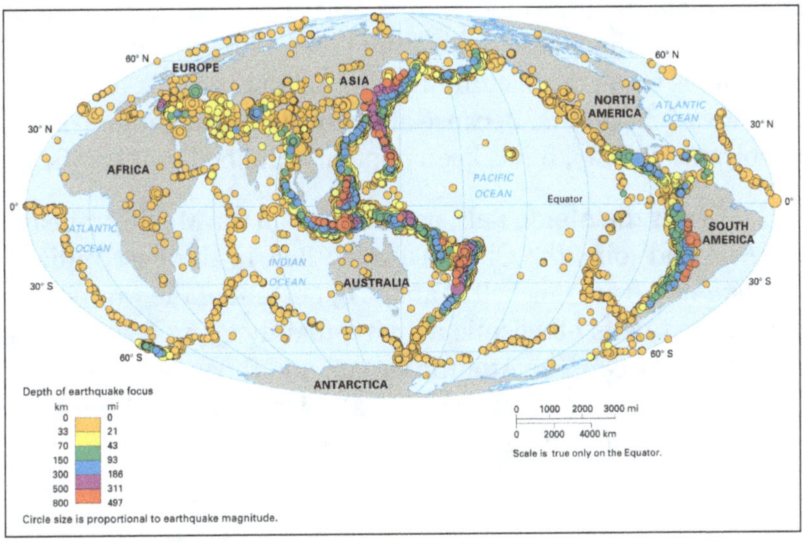

Global seismic centres in 1975-99: Earthquakes of magnitude 5.5 and greater.

About 50,000 earthquakes large enough to be noticed without the aid of instruments occur annually over the entire Earth. Of these, approximately 100 are of sufficient size to produce substantial damage if their centres are near areas of habitation. Very great earthquakes occur on average about once per year. Over the centuries they have been responsible for millions of deaths and an incalculable amount of damage to property.

Crowds watching the fires set off by the earthquake in San Francisco in 1906.

The Nature of Earthquakes

Causes of Earthquakes

Earth's major earthquakes occur mainly in belts coinciding with the margins of tectonic plates. This has long been apparent from early catalogs of felt earthquakes and is even more readily discernible in modern seismicity maps, which show instrumentally determined epicentres. The most important earthquake belt is the Circum-Pacific Belt, which affects many populated coastal regions around the Pacific Ocean—for example, those of New Zealand, New Guinea, Japan, the Aleutian Islands, Alaska, and the western coasts of North and South America. It is estimated that 80 percent of the energy presently released in earthquakes comes from those whose epicentres are in this belt. The seismic activity is by no means uniform throughout the belt, and there are a number of branches at various points. Because at many places the Circum-Pacific Belt is associated with volcanic activity, it has been popularly dubbed the "Pacific Ring of Fire."

A second belt, known as the Alpide Belt, passes through the Mediterranean region eastward through Asia and joins the Circum-Pacific Belt in the East Indies. The energy released in earthquakes from this belt is about 15 percent of the world total. There also are striking connected belts of seismic activity, mainly along oceanic ridges—including those in the Arctic Ocean, the Atlantic Ocean, and the western Indian Ocean—and along the rift valleys of East Africa. This global seismicity distribution is best understood in terms of its plate tectonic setting.

Natural Forces

Earthquakes are caused by the sudden release of energy within some limited region of the rocks of the Earth. The energy can be released by elastic strain, gravity, chemical reactions, or even the motion of massive bodies. Of all these the release of elastic strain is the most important cause, because this form of energy is the only kind that can be stored in sufficient quantity in the Earth to produce major disturbances. Earthquakes associated with this type of energy release are called tectonic earthquakes.

Tectonics

Tectonic earthquakes are explained by the so-called elastic rebound theory, formulated by the American geologist Harry Fielding Reid after the San Andreas Fault ruptured in 1906, generating the great San Francisco earthquake. According to the theory, a tectonic earthquake occurs when strains in rock masses have accumulated to a point where the resulting stresses exceed the strength of the rocks, and sudden fracturing results. The fractures propagate rapidly through the rock, usually tending in the same direction and sometimes extending many kilometres along a local zone of weakness. In 1906, for instance, the San Andreas Fault slipped along a plane 430 km (270 miles) long. Along this line the ground was displaced horizontally as much as 6 metres (20 feet).

As a fault rupture progresses along or up the fault, rock masses are flung in opposite directions and thus spring back to a position where there is less strain. At any one point this movement may take place not at once but rather in irregular steps; these sudden slowings and restartings give rise to the vibrations that propagate as seismic waves. Such irregular properties of fault rupture are now included in the modeling of earthquake sources, both physically and mathematically. Roughnesses along the fault are referred to as asperities, and places where the rupture slows or stops are said to be fault barriers. Fault rupture starts at the earthquake focus, a spot that in many cases is close to 5–15 km under the surface. The rupture propagates in one or both directions over the fault plane until stopped or slowed at a barrier. Sometimes, instead of being stopped at the barrier, the fault rupture recommences on the far side; at other times the stresses in the rocks break the barrier, and the rupture continues.

Earthquakes have different properties depending on the type of fault slip that causes them. The usual fault model has a "strike" (that is, the direction from north taken by a horizontal line in the fault plane) and a "dip" (the angle from the horizontal shown by the steepest slope in the fault). The lower wall of an inclined fault is called the footwall. Lying over the footwall is the hanging wall. When rock masses slip past each other parallel to the strike, the movement is known as strike-slip faulting. Movement parallel to the dip is called dip-slip faulting. Strike-slip faults are right lateral or left lateral, depending on whether the block on the opposite side of the fault from an observer has moved to the right or left. In dip-slip faults, if the hanging-wall block moves downward relative to the footwall block, it is called "normal" faulting; the opposite motion, with the hanging wall moving upward relative to the footwall, produces reverse or thrust faulting.

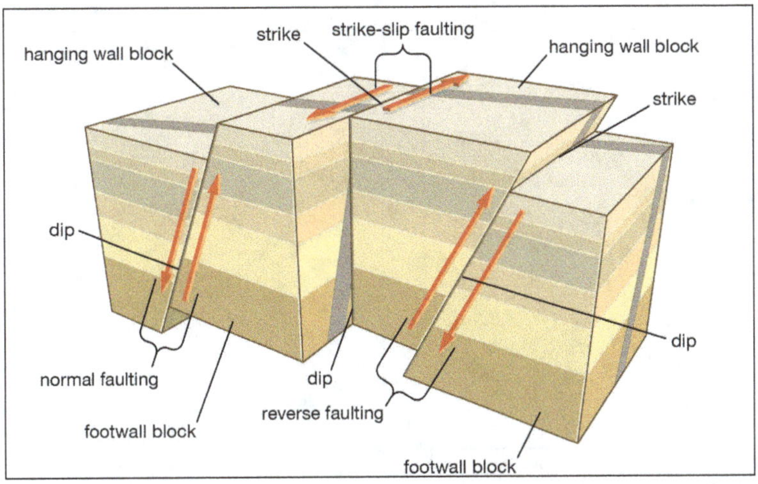

Types of faulting in tectonic earthquakes: In normal and reverse faulting, rock masses slip vertically past each other. In strike-slip faulting, the rocks slip past each other horizontally.

All known faults are assumed to have been the seat of one or more earthquakes in the past, though tectonic movements along faults are often slow, and most geologically ancient faults are now aseismic (that is, they no longer cause earthquakes). The actual faulting associated with an earthquake may be complex, and it is often not clear whether in a particular earthquake the total energy issues from a single fault plane.

Observed geologic faults sometimes show relative displacements on the order of hundreds of kilometres over geologic time, whereas the sudden slip offsets that produce seismic waves may range from only several centimetres to tens of metres. In the 1976 Tangshan earthquake, for example, a surface strike-slip of about one metre was observed along the causative fault east of Beijing, and in the 1999 Taiwan earthquake the Chelung-pu fault slipped up to eight metres vertically.

Volcanism

A separate type of earthquake is associated with volcanic activity and is called a volcanic earthquake. Yet it is likely that even in such cases the disturbance is the result of a sudden slip of rock masses adjacent to the volcano and the consequent release of elastic strain energy. The stored energy, however, may in part be of hydrodynamic origin due to heat provided by magma moving in reservoirs beneath the volcano or to the release of gas under pressure.

There is a clear correspondence between the geographic distribution of volcanoes and major earthquakes, particularly in the Circum-Pacific Belt and along oceanic ridges. Volcanic vents, however, are generally several hundred kilometres from the epicentres of most major shallow earthquakes, and many earthquake sources occur nowhere near active volcanoes. Even in cases where an earthquake's focus occurs directly below structures marked by volcanic vents, there is probably no immediate causal connection between the two activities; most likely both are the result of the same tectonic processes.

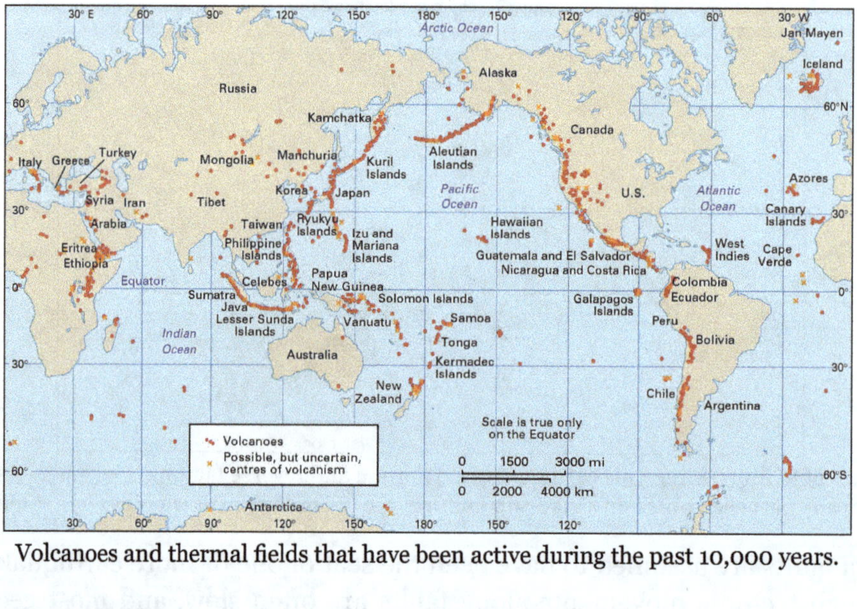

Volcanoes and thermal fields that have been active during the past 10,000 years.

Artificial Induction

Earthquakes are sometimes caused by human activities, including the injection of fluids into deep wells, the detonation of large underground nuclear explosions, the excavation of mines, and the filling of large reservoirs. In the case of deep mining, the removal of rock produces changes in the strain around the tunnels. Slip on adjacent, preexisting faults or outward shattering of rock into the new cavities may occur. In fluid injection, the slip is thought to be induced by premature release of elastic strain, as in the case of tectonic earthquakes, after fault surfaces are lubricated by the liquid. Large underground nuclear explosions have been known to produce slip on already strained faults in the vicinity of the test devices.

Reservoir Induction

Of the various earthquake-causing activities cited above, the filling of large reservoirs is among the most important. More than 20 significant cases have been documented in which local seismicity has increased following the impounding of water behind high dams. Often, causality cannot be substantiated, because no data exists to allow comparison of earthquake occurrence before and after the reservoir was filled. Reservoir-induction effects are most marked for reservoirs exceeding 100 metres (330 feet) in depth and 1 cubic km (0.24 cubic mile) in volume. Three sites where such connections have very probably occurred are the Hoover Dam in the United States, the Aswan High Dam in Egypt, and the Kariba Dam on the border between Zimbabwe and Zambia. The most generally accepted explanation for earthquake occurrence in such cases assumes that rocks near the reservoir are already strained from regional tectonic forces to a point where nearby faults are almost ready to slip. Water in the reservoir adds a pressure perturbation that triggers the fault rupture. The pressure effect is perhaps enhanced by the fact that the rocks along the fault have lower strength because of increased water-pore pressure. These factors notwithstanding, the filling of most large reservoirs has not produced earthquakes large enough to be a hazard.

The specific seismic source mechanisms associated with reservoir induction have been established in a few cases. For the main shock at the Koyna Dam and Reservoir in India, the evidence favours strike-slip faulting motion. At both the Kremasta Dam in Greece and the Kariba Dam in Zimbabwe-Zambia, the generating mechanism was dip-slip on normal faults. By contrast, thrust mechanisms have been determined for sources of earthquakes at the lake behind Nurek Dam in Tajikistan. More than 1,800 earthquakes occurred during the first nine years after water was impounded in this 317-metre-deep reservoir in 1972, a rate amounting to four times the average number of shocks in the region prior to filling.

Seismology and Nuclear Explosions

In 1958 representatives from several countries, including the United States and the Soviet Union, met to discuss the technical basis for a nuclear test-ban treaty. Among the matters considered was the feasibility of developing effective means with which to detect underground nuclear explosions and to distinguish them seismically from earthquakes. After that conference, much special research was directed to seismology, leading to major advances in seismic signal detection and analysis.

Recent seismological work on treaty verification has involved using high-resolution seismographs in a worldwide network, estimating the yield of explosions, studying wave attenuation in the Earth, determining wave amplitude and frequency spectra discriminants, and applying seismic arrays. The findings of such research have shown that underground nuclear explosions, compared with natural earthquakes, usually generate seismic waves through the body of the Earth that are of much larger amplitude than the surface waves. This tell-tale difference along with other types of seismic evidence suggest that an international monitoring network of 270 seismographic stations could detect and locate all seismic events over the globe of magnitude 4 and above (corresponding to an explosive yield of about 100 tons of TNT).

Earthquake Preparedness

Earthquake preparedness is a set of measures taken at the individual, organisational and societal level to minimise the effects of an earthquake. Preparedness measures can range from securing heavy objects, structural modifications and storing supplies, to having insurance, an emergency kit, and evacuation plans.

Preparedness Measures

Preparedness can consist of survival measures, preparation that will improve survival in the event of an earthquake, or mitigating measures, that seek to minimise the effect of an earthquake. Common survival measures include storing food and water for an emergency, and educating individuals what to do during an earthquake. Mitigating measures can include firmly securing large items of furniture (such as bookcases and large cabinets), TV and computer screens that may otherwise fall over in an earthquake. Likewise, avoiding storing items above beds or sofas reduces the chance of objects falling on individuals. Planning for a related tsunami, tsunami preparedness, can also be part of earthquake preparedness.

Building Design and Retrofitting

Building codes in earthquake prone areas may have specific requirements designed to increase new buildings' resistance to earthquakes. Older buildings and homes that are not up to code may be modified to increase their resistance. Modification and earthquake resistant design are also employed in elevated freeways and bridges.

Codes are not designed to make buildings earthquake proof in the sense of them suffering zero damage. The goal of most building designs is to reduce earthquake damage to a building such that it protects the lives of occupants and thus tolerance of some limited damage is accepted and considered a necessary tradeoff. A supplement or precursor to retrofitting can be the implementation of earthquake proof furniture.

Earthquake modification techniques and modern building codes are designed to prevent total destruction of buildings for earthquakes of no greater than 8.5 on the Richter Scale. Although the Richter Scale is referenced, the localized shaking intensity is one of the largest factors to be considered in building resiliency.

Types of Preparedness

The basic theme behind preparedness is to be ready for an earthquake. Preparedness starts with an individual's everyday life and involves items and training that would be useful in an earthquake. Preparedness continues on a continuum from individual preparedness through family preparedness, community preparedness and then business, non-profit and governmental preparedness. Some organisations blend these various levels. Business continuity planning encourages businesses to have a Disaster Recovery Plan. The US FEMA breaks down preparedness generally into a pyramid, with citizens on the foundational bottom, on top of which rests local government, state government and federal government in that order.

Children may present particular issues and some planning and resources are directly focused on supporting them. The US FEMA has advice noting that "Disasters can leave children feeling frightened, confused, and insecure" whether a child has experienced it first hand, had it happen to a friend or simply seen it on television. People with disabilities or other special needs may have special emergency preparation needs. FEMA's suggestions for people with disabilities include having copies of prescriptions, charging devices for medical devices such as motorized wheel chairs and a week's supply of medication readily available. Preparedness can also cover pets.

Preparedness can also encompass psychological preparedness: resources are designed to support both community members affected by a disaster and the disaster workers serving them.

A multi-hazard approach, where communities are prepared for several hazards, are more resilient than single hazard approaches and have been gaining popularity.

Non-perishable food in cabinet.

Long term power outages can cause damage beyond the original disaster that can be mitigated with emergency generators or other power sources to provide an emergency power system. The United States Department of Energy states: "homeowners, business owners, and local leaders may have to take an active role in dealing with energy disruptions on their own." Major institutions like hospitals, military bases and educational institutions often have extensive backup power systems. Preparedness does not stop at home or at school. The United States Department of Health and Human Services addresses specific emergency preparedness issues hospitals may have to respond to, including maintaining a safe temperature, providing adequate electricity for life support systems and even carrying out evacuations under extreme circumstances. FEMA encourages all businesses to have an emergency response plan and the Small Business Administration specifically advises small business owners to also focus emergency preparedness and provides a variety of different worksheets and resources.

Marked gas shutoff.

Given the explosive danger posed by natural gas leaks, Ready.gov states that "It is vital that all household members know how to shut off natural gas" and that property owners must ensure they have any special tools needed for their particular gas connections. Ready.gov also notes that "It is wise to teach all responsible household members where and how to shut off the electricity," cautioning that individual circuits should be shut off before the main circuit. Ready.gov further states that "It is vital that all household members learn how to shut off the water at the main house valve" and cautions that the possibility that rusty valves might require replacement.

Achieving Preparedness

Levels of preparedness generally remain low, despite attempts to increase public awareness.

Various methods exist to promote disaster preparedness, but they are rarely well documented and their efficacy is rarely tested. Hands on training, drills and face-to-face interaction have proven more successful at changing behaviour. Digital methods have also been used, including for examples educational videogames.

Flood

A flood is an overflow of water that submerges land that is usually dry. In the sense of "flowing water", the word may also be applied to the inflow of the tide. Floods are an area of study of the discipline hydrology and are of significant concern in agriculture, civil engineering and public health.

Flooding in a street.

Flooding may occur as an overflow of water from water bodies, such as a river, lake, or ocean, in which the water overtops or breaks levees, resulting in some of that water escaping its usual boundaries, or it may occur due to an accumulation of rainwater on saturated ground in an areal flood. While the size of a lake or other body of water will vary with seasonal changes in precipitation and snow melt, these changes in size are unlikely to be considered significant unless they flood property or drown domestic animals.

View of flooded New Orleans in the aftermath of Hurricane Katrina.

Floods can also occur in rivers when the flow rate exceeds the capacity of the river channel, particularly at bends or meanders in the waterway. Floods often cause damage to homes and businesses if they are in the natural flood plains of rivers. While riverine flood damage can be eliminated by moving away from rivers and other bodies of water, people have traditionally lived and worked by rivers because the land is usually flat and fertile and because rivers provide easy travel and access to commerce and industry.

Some floods develop slowly, while others can develop in just a few minutes and without visible signs of rain. Additionally, floods can be local, impacting a neighborhood or community, or very large, affecting entire river basins.

"Regular" flooding in Venice, Italy.

Causes

Upslope Factors

Flood due to Cyclone Hudhud in Visakhapatnam.

The amount, location, and timing of water reaching a drainage channel from natural precipitation and controlled or uncontrolled reservoir releases determines the flow at downstream locations. Some precipitation evaporates, some slowly percolates through soil, some may be temporarily sequestered as snow or ice, and some may produce rapid runoff from surfaces including rock, pavement, roofs, and saturated or frozen ground. The fraction of incident precipitation promptly reaching a drainage channel has been observed from nil for light rain on dry, level ground to as high as 170 percent for warm rain on accumulated snow.

Most precipitation records are based on a measured depth of water received within a fixed time interval. Frequency of a precipitation threshold of interest may be determined from the number of measurements exceeding that threshold value within the total time period for which observations are available. Individual data points are converted to intensity by dividing each measured depth by the period of time between observations. This intensity will be less than the actual peak intensity if the duration of the rainfall event was less than the fixed time interval for which measurements are reported. Convective precipitation events (thunderstorms) tend to produce shorter duration storm events than orographic precipitation. Duration, intensity, and frequency of rainfall events are important to flood prediction. Short duration precipitation is more significant to flooding within small drainage basins.

The most important upslope factor in determining flood magnitude is the land area of the watershed upstream of the area of interest. Rainfall intensity is the second most important factor for watersheds of less than approximately 30 square miles or 80 square kilometres. The main channel slope is the second most important factor for larger watersheds. Channel slope and rainfall intensity become the third most important factors for small and large watersheds, respectively.

Time of Concentration is the time required for runoff from the most distant point of the upstream drainage area to reach the point of the drainage channel controlling flooding of the area of interest. The time of concentration defines the critical duration of peak rainfall for the area of interest. The critical duration of intense rainfall might be only a few minutes for roof and parking lot drainage structures, while cumulative rainfall over several days would be critical for river basins.

Downslope Factors

Water flowing downhill ultimately encounters downstream conditions slowing movement. The final limitation in coastal flooding lands is often the ocean or some coastal flooding bars which form natural lakes. In flooding low lands, elevation changes such as tidal fluctuations are significant determinants of coastal and estuarine flooding. Less predictable events like tsunamis and storm surges may also cause elevation changes in large bodies of water. Elevation of flowing water is controlled by the geometry of the flow channel and, especially, by depth of channel, speed of flow and amount of sediments in it Flow channel restrictions like bridges and canyons tend to control water elevation above the restriction. The actual control point for any given reach of the drainage may change with changing water elevation, so a closer point may control for lower water levels until a more distant point controls at higher water levels.

Effective flood channel geometry may be changed by growth of vegetation, accumulation of ice or debris, or construction of bridges, buildings, or levees within the flood channel.

Coincidence

Extreme flood events often result from coincidence such as unusually intense, warm rainfall melting heavy snow pack, producing channel obstructions from floating ice, and releasing small impoundments like beaver dams. Coincident events may cause extensive flooding to be more frequent than anticipated from simplistic statistical prediction models considering only precipitation runoff flowing within unobstructed drainage channels. Debris modification of channel geometry is common when heavy flows move uprooted woody vegetation and flood-damaged structures and vehicles, including boats and railway equipment. Recent field measurements during the 2010–11 Queensland floods showed that any criterion solely based upon the flow velocity, water depth or specific momentum cannot account for the hazards caused by velocity and water depth fluctuations. These considerations ignore further the risks associated with large debris entrained by the flow motion.

Some researchers have mentioned the storage effect in urban areas with transportation corridors created by cut and fill. Culverted fills may be converted to impoundments if the culverts become blocked by debris, and flow may be diverted along streets. Several studies have looked into the flow patterns and redistribution in streets during storm events and the implication on flood modelling.

Flood Safety Planning

Aftermath of flooding in Colorado, 2013.

In the United States, the National Weather Service gives out the advice Turn Around, Don't Drown" for floods; that is, it recommends that people get out of the area of a flood, rather than trying to cross it. At the most basic level, the best defense against floods is to seek higher ground for high-value uses while balancing the foreseeable risks with the benefits of occupying flood hazard zones. Critical community-safety facilities, such as hospitals, emergency-operations centers, and police, fire, and rescue services, should be built in areas least at risk of flooding. Structures, such as bridges, that must unavoidably be in flood hazard areas should be designed to withstand flooding. Areas most at risk for flooding could be put to valuable uses that could be abandoned temporarily as people retreat to safer areas when a flood is imminent.

Planning for flood safety involves many aspects of analysis and engineering, including:

- Observation of previous and present flood heights and inundated areas.
- Statistical, hydrologic, and hydraulic model analyses.
- Mapping inundated areas and flood heights for future flood scenarios.
- Long-term land use planning and regulation.
- Engineering design and construction of structures to control or withstand flooding.
- Intermediate-term monitoring, forecasting, and emergency-response planning.
- Short-term monitoring, warning, and response operations.

Each topic presents distinct yet related questions with varying scope and scale in time, space, and the people involved. Attempts to understand and manage the mechanisms at work in floodplains have been made for at least six millennia.

In the United States, the Association of State Floodplain Managers works to promote education, policies, and activities that mitigate current and future losses, costs, and human suffering caused by flooding and to protect the natural and beneficial functions of floodplains – all without causing adverse impacts. A portfolio of best practice examples for disaster mitigation in the United States is available from the Federal Emergency Management Agency.

Control

In many countries around the world, waterways prone to floods are often carefully managed. Defenses such as detention basins, levees, bunds, reservoirs, and weirs are used to prevent waterways from overflowing their banks. When these defenses fail, emergency measures such as sandbags or portable inflatable tubes are often used to try to stem flooding. Coastal flooding has been addressed in portions of Europe and the Americas with coastal defenses, such as sea walls, beach nourishment, and barrier islands.

In the riparian zone near rivers and streams, erosion control measures can be taken to try to slow down or reverse the natural forces that cause many waterways to meander over long periods of time. Flood controls, such as dams, can be built and maintained over time to try to reduce the occurrence and severity of floods as well. In the United States, the U.S. Army Corps of Engineers maintains a network of such flood control dams.

In areas prone to urban flooding, one solution is the repair and expansion of man-made sewer systems and stormwater infrastructure. Another strategy is to reduce impervious surfaces in streets, parking lots and buildings through natural drainage channels, porous paving, and wetlands (collectively known as green infrastructure or sustainable urban drainage systems (SUDS)). Areas identified as flood-prone can be converted into parks and playgrounds that can tolerate occasional flooding. Ordinances can be adopted to require developers to retain stormwater on site and require buildings to be elevated, protected by floodwalls and levees, or designed to withstand temporary inundation. Property owners can also invest in solutions themselves, such as re-landscaping their property to take the flow of water away from their building and installing rain barrels, sump pumps, and check valves.

Analysis of Flood Information

A series of annual maximum flow rates in a stream reach can be analyzed statistically to estimate the 100-year flood and floods of other recurrence intervals there. Similar estimates from many sites in a hydrologically similar region can be related to measurable characteristics of each drainage basin to allow indirect estimation of flood recurrence intervals for stream reaches without sufficient data for direct analysis.

Physical process models of channel reaches are generally well understood and will calculate the depth and area of inundation for given channel conditions and a specified flow rate, such as for use in floodplain mapping and flood insurance. Conversely, given the observed inundation area of a recent flood and the channel conditions, a model can calculate the flow rate. Applied to various potential channel configurations and flow rates, a reach model can contribute to selecting an optimum design for a modified channel. Various reach models are available as of 2015, either 1D models (flood levels measured in the channel) or 2D models (variable flood depths measured across the extent of a floodplain). HEC-RAS, the Hydraulic Engineering Center model, is among the most popular software, if only because it is available free of charge. Other models such as TUFLOW combine 1D and 2D components to derive flood depths across both river channels and the entire floodplain.

Physical process models of complete drainage basins are even more complex. Although many processes are well understood at a point or for a small area, others are poorly understood at all scales, and process interactions under normal or extreme climatic conditions may be unknown. Basin models typically combine land-surface process components (to estimate how much rainfall or snowmelt reaches a channel) with a series of reach models. For example, a basin model can calculate the runoff hydrograph that might result from a 100-year storm, although the recurrence interval of a storm is rarely equal to that of the associated flood. Basin models are commonly used in flood forecasting and warning, as well as in analysis of the effects of land use change and climate change.

Flood Forecasting

Anticipating floods before they occur allows for precautions to be taken and people to be warned so that they can be prepared in advance for flooding conditions. For example, farmers can remove animals from low-lying areas and utility services can put in place emergency provisions to re-route services if needed. Emergency services can also make provisions to have enough resources available ahead of time to respond to emergencies as they occur. People can evacuate areas to be flooded.

In order to make the most accurate flood forecasts for waterways, it is best to have a long time-series of historical data that relates stream flows to measured past rainfall events. Coupling this historical information with real-time knowledge about volumetric capacity in catchment areas, such as spare capacity in reservoirs, ground-water levels, and the degree of saturation of area aquifers is also needed in order to make the most accurate flood forecasts.

Radar estimates of rainfall and general weather forecasting techniques are also important components of good flood forecasting. In areas where good quality data is available, the intensity and height of a flood can be predicted with fairly good accuracy and plenty of lead time. The output of a flood forecast is typically a maximum expected water level and the likely time of its arrival at key locations along a waterway, and it also may allow for the computation of the likely statistical return period of a flood. In many developed countries, urban areas at risk of flooding are protected against a 100-year flood – that is a flood that has a probability of around 63% of occurring in any 100-year period of time.

According to the U.S. National Weather Service (NWS) Northeast River Forecast Center (RFC) in Taunton, Massachusetts, a rule of thumb for flood forecasting in urban areas is that it takes at least 1 inch (25 mm) of rainfall in around an hour's time in order to start significant ponding of water on impermeable surfaces. Many NWS RFCs routinely issue Flash Flood Guidance and Headwater Guidance, which indicate the general amount of rainfall that would need to fall in a short period of time in order to cause flash flooding or flooding on larger water basins.

In the United States, an integrated approach to real-time hydrologic computer modelling utilizes observed data from the U.S. Geological Survey (USGS), various cooperative observing networks, various automated weather sensors, the NOAA National Operational Hydrologic Remote Sensing Center (NOHRSC), various hydroelectric companies, etc. combined with quantitative precipitation forecasts (QPF) of expected rainfall and/or snow melt to generate daily or as-needed hydrologic forecasts. The NWS also cooperates with Environment Canada on hydrologic forecasts that affect both the US and Canada, like in the area of the Saint Lawrence Seaway.

The Global Flood Monitoring System, "GFMS," a computer tool which maps flood conditions worldwide, is available online. Users anywhere in the world can use GFMS to determine when floods may occur in their area. GFMS uses precipitation data from NASA's Earth observing satellites and the Global Precipitation Measurement satellite, "GPM." Rainfall data from GPM is combined with a land surface model that incorporates vegetation cover, soil type, and terrain to determine how much water is soaking into the ground, and how much water is flowing into streamflow.

Users can view statistics for rainfall, streamflow, water depth, and flooding every 3 hours, at each 12 kilometer gridpoint on a global map. Forecasts for these parameters are 5 days into the future. Users can zoom in to see inundation maps (areas estimated to be covered with water) in 1 kilometer resolution.

Tsunami

A tsunami is a series of waves in a water body caused by the displacement of a large volume of water, generally in an ocean or a large lake. Earthquakes, volcanic eruptions and other underwater explosions (including detonations, landslides, glacier calvings, meteorite impacts and other disturbances) above or below water all have the potential to generate a tsunami. Unlike normal ocean waves, which are generated by wind, or tides, which are generated by the gravitational pull of the Moon and the Sun, a tsunami is generated by the displacement of water.

The 2004 Indian Ocean tsunami at Ao Nang, Krabi Province, Thailand.

Tsunami waves do not resemble normal undersea currents or sea waves because their wavelength is far longer. Rather than appearing as a breaking wave, a tsunami may instead initially resemble a rapidly rising tide. For this reason, it is often referred to as a tidal wave, although this usage is not favoured by the scientific community because it might give the false impression of a causal relationship between tides and tsunamis. Tsunamis generally consist of a series of waves, with periods ranging from minutes to hours, arriving in a so-called "wave train". Wave heights of tens of metres can be generated by large events. Although the impact of tsunamis is limited to coastal areas, their destructive power can be enormous, and they can affect entire ocean basins. The 2004 Indian Ocean tsunami was among the deadliest natural disasters in human history, with at least 230,000 people killed or missing in 14 countries bordering the Indian Ocean.

The Ancient Greek historian Thucydides suggested in his 5th century BC *History of the Peloponnesian War* that tsunamis were related to submarine earthquakes, but the understanding of tsunamis remained slim until the 20th century and much remains unknown. Major areas of current research include determining why some large earthquakes do not generate tsunamis while other smaller ones do; accurately forecasting the passage of tsunamis across the oceans; and forecasting how tsunami waves interact with shorelines.

Causes

The principal generation mechanism of a tsunami is the displacement of a substantial volume of water or perturbation of the sea. This displacement of water is usually attributed to either earthquakes, landslides, volcanic eruptions, glacier calvings or more rarely by meteorites and nuclear tests.

Seismicity

Tsunamis can be generated when the sea floor abruptly deforms and vertically displaces the overlying water. Tectonic earthquakes are a particular kind of earthquake that are associated with the Earth's crustal deformation; when these earthquakes occur beneath the sea, the water above the deformed area is displaced from its equilibrium position. More specifically, a tsunami can be generated when thrust faults associated with convergent or destructive plate boundaries move abruptly, resulting in water displacement, owing to the vertical component of movement involved. Movement on normal (extensional) faults can also cause displacement of the seabed, but only the largest of such events (typically related to flexure in the outer trench swell) cause enough displacement to give rise to a significant tsunami, such as the 1977 Sumba and 1933 Sanriku events.

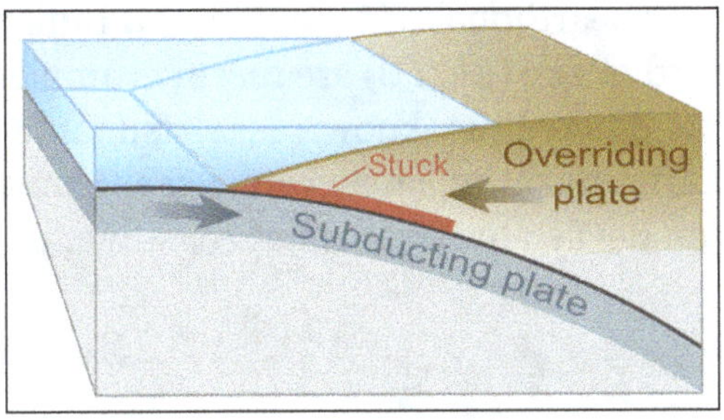

Drawing of tectonic plate boundary before earthquake.

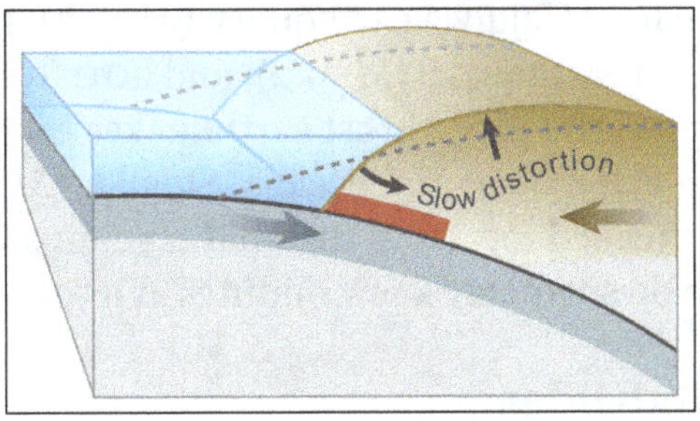

Over-riding plate bulges under strain, causing tectonic uplift.

Tsunamis have a small wave height offshore, and a very long wavelength (often hundreds of kilometres long, whereas normal ocean waves have a wavelength of only 30 or 40 metres), which is why they generally pass unnoticed at sea, forming only a slight swell usually about 300 millimetres (12 in) above the normal sea surface. They grow in height when they reach shallower water, in a wave shoaling process described below. A tsunami can occur in any tidal state and even at low tide can still inundate coastal areas.

On April 1, 1946, the 8.6 M_w Aleutian Islands earthquake occurred with a maximum Mercalli intensity of VI (*Strong*). It generated a tsunami which inundated Hilo on the island of Hawaii with a 14-metre high (46 ft) surge. Between 165 and 173 were killed.

The area where the earthquake occurred is where the Pacific Ocean floor is subducting (or being pushed downwards) under Alaska.

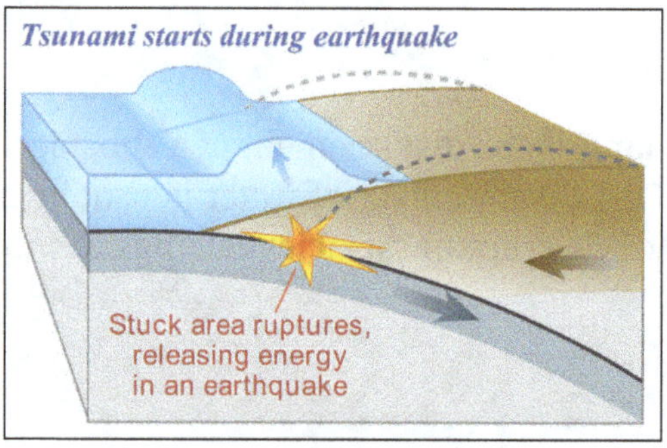

Plate slips, causing subsidence and releasing energy into water.

The energy released produces tsunami waves.

Examples of tsunamis originating at locations away from convergent boundaries include Storegga about 8,000 years ago, Grand Banks in 1929, and Papua New Guinea in 1998. The Grand Banks and Papua New Guinea tsunamis came from earthquakes which destabilised sediments, causing them to flow into the ocean and generate a tsunami. They dissipated before travelling transoceanic distances.

The cause of the Storegga sediment failure is unknown. Possibilities include an overloading of the sediments, an earthquake or a release of gas hydrates (methane etc.).

The 1960 Valdivia earthquake (M_w 9.5), 1964 Alaska earthquake (M_w 9.2), 2004 Indian Ocean earthquake (M_w 9.2), and 2011 Tōhoku earthquake (M_w9.0) are recent examples of powerful megathrust earthquakes that generated tsunamis (known as teletsunamis) that can cross entire oceans. Smaller (M_w 4.2) earthquakes in Japan can trigger tsunamis (called local and regional tsunamis) that can devastate stretches of coastline, but can do so in only a few minutes at a time.

Landslides

In the 1950s, it was discovered that tsunamis larger than had previously been believed possible can be caused by giant submarine landslides. These rapidly displace large water volumes, as energy transfers to the water at a rate faster than the water can absorb. Their existence was confirmed in 1958, when a giant landslide in Lituya Bay, Alaska, caused the highest wave ever recorded, which had a height of 524 metres (over 1700 feet). The wave did not travel far, as it struck land almost immediately. Two people fishing in the bay were killed, but another boat managed to ride the wave.

Another landslide-tsunami event occurred in 1963 when a massive landslide from Monte Toc entered the Vajont Dam in Italy. The resulting wave surged over the 262 m (860 ft) high dam by 250 metres (820 ft) and destroyed several towns. Around 2,000 people died. Scientists named these waves megatsunamis.

Some geologists claim that large landslides from volcanic islands, e.g. Cumbre Vieja on La Palma in the Canary Islands, may be able to generate megatsunamis that can cross oceans, but this is disputed by many others.

In general, landslides generate displacements mainly in the shallower parts of the coastline, and there is conjecture about the nature of large landslides that enter the water. This has been shown to subsequently affect water in enclosed bays and lakes, but a landslide large enough to cause a transoceanic tsunami has not occurred within recorded history. Susceptible locations are believed to be the Big Island of Hawaii, Fogo in the Cape Verde Islands, La Reunion in the Indian Ocean, and Cumbre Vieja on the island of La Palma in the Canary Islands; along with other volcanic ocean islands. This is because large masses of relatively unconsolidated volcanic material occurs on the flanks and in some cases detachment planes are believed to be developing. However, there is growing controversy about how dangerous these slopes actually are.

Meteorological

Some meteorological conditions, especially rapid changes in barometric pressure, as seen with the passing of a front, can displace bodies of water enough to cause trains of waves with wavelengths comparable to seismic tsunamis, but usually with lower energies. These are essentially dynamically equivalent to seismic tsunamis, the only differences being that meteotsunamis lack the transoceanic reach of significant seismic tsunamis and that the force that displaces the water is sustained over some length of time such that meteotsunamis cannot be modelled as having been caused instantaneously. In spite of their lower energies, on shorelines where they can be amplified by resonance, they are sometimes powerful enough to cause localised damage and potential for loss of life. They have been documented in many places, including the Great Lakes, the Aegean Sea, the English Channel, and the Balearic Islands, where they are common enough to have a local name, *rissaga*. In Sicily they are called *marubbio* and in Nagasaki Bay, they are called *abiki*. Some examples of destructive meteotsunamis include 31 March 1979 at Nagasaki and 15 June 2006 at Menorca, the latter causing damage in the tens of millions of euros.

Meteotsunamis should not be confused with storm surges, which are local increases in sea level associated with the low barometric pressure of passing tropical cyclones, nor should they be confused with setup, the temporary local raising of sea level caused by strong on-shore winds. Storm surges and setup are also dangerous causes of coastal flooding in severe weather but their dynamics are completely unrelated to tsunami waves. They are unable to propagate beyond their sources, as waves do.

Man-made or Triggered Tsunamis

There have been studies of the potential of the induction of and at least one actual attempt to create tsunami waves as a tectonic weapon.

In World War II, the New Zealand Military Forces initiated Project Seal, which attempted to create small tsunamis with explosives in the area of today's Shakespear Regional Park; the attempt failed.

There has been considerable speculation on the possibility of using nuclear weapons to cause tsunamis near an enemy coastline. Even during World War II consideration of the idea using conventional explosives was explored. Nuclear testing in the Pacific Proving Ground by the United States seemed to generate poor results. *Operation Crossroads* fired two 20 kilotonnes of TNT (84 TJ) bombs, one in the air and one underwater, above and below the shallow (50 m (160 ft)) waters of the Bikini Atoll lagoon. Fired about 6 km (3.7 mi) from the nearest island, the waves there were no higher than 3–4 m (9.8–13.1 ft) upon reaching the shoreline. Other underwater tests, mainly *Hardtack I/Wahoo* (deep water) and *Hardtack I/Umbrella* (shallow water) confirmed the results. Analysis of the effects of shallow and deep underwater explosions indicate that the energy of the explosions does not easily generate the kind of deep, all-ocean waveforms which are tsunamis; most of the energy creates steam, causes vertical fountains above the water, and creates compressional waveforms. Tsunamis are hallmarked by permanent large vertical displacements of very large volumes of water which do not occur in explosions.

Characteristics

Tsunamis cause damage by two mechanisms: the smashing force of a wall of water travelling at high speed, and the destructive power of a large volume of water draining off the land and carrying a large amount of debris with it, even with waves that do not appear to be large.

While everyday wind waves have a wavelength (from crest to crest) of about 100 metres (330 ft) and a height of roughly 2 metres (6.6 ft), a tsunami in the deep ocean has a much larger wavelength of up to 200 kilometres (120 mi). Such a wave travels at well over 800 kilometres per hour (500 mph), but owing to the enormous wavelength the wave oscillation at any given point takes 20 or 30 minutes to complete a cycle and has an amplitude of only about 1 metre (3.3 ft). This makes tsunamis difficult to detect over deep water, where ships are unable to feel their passage.

The wave further slows and amplifies as it hits land. Only the largest waves crest.

The velocity of a tsunami can be calculated by obtaining the square root of the depth of the water in metres multiplied by the acceleration due to gravity (approximated to 10 m/s^2). For example, if the Pacific Ocean is considered to have a depth of 5000 metres, the velocity of a tsunami would be the square root of $\sqrt{(5000 \times 10)} = \sqrt{50000} = \sim 224$ metres per second (735 feet per second), which equates to a speed of ~806 kilometres per hour or about 500 miles per hour. This is the formula used for calculating the velocity of shallow-water waves. Even the deep ocean is shallow in this sense because a tsunami wave is so long (horizontally from crest to crest) by comparison.

The reason for the Japanese name "harbour wave" is that sometimes a village's fishermen would sail out, and encounter no unusual waves while out at sea fishing, and come back to land to find their village devastated by a huge wave.

As the tsunami approaches the coast and the waters become shallow, wave shoaling compresses the wave and its speed decreases below 80 kilometres per hour (50 mph). Its wavelength diminishes to less than 20 kilometres (12 mi) and its amplitude grows enormously – in accord with Green's law. Since the wave still has the same very long period, the tsunami may take minutes to reach full height. Except for the very largest tsunamis, the approaching wave does not break, but rather appears like a fast-moving tidal bore. Open bays and coastlines adjacent to very deep water may shape the tsunami further into a step-like wave with a steep-breaking front.

When the tsunami's wave peak reaches the shore, the resulting temporary rise in sea level is termed *run up*. Run up is measured in metres above a reference sea level. A large tsunami may feature multiple waves arriving over a period of hours, with significant time between the wave crests. The first wave to reach the shore may not have the highest run-up.

About 80% of tsunamis occur in the Pacific Ocean, but they are possible wherever there are large bodies of water, including lakes. They are caused by earthquakes, landslides, volcanic explosions, glacier calvings, and bolides.

Drawback

All waves have a positive and negative peak; that is, a ridge and a trough. In the case of a propagating wave like a tsunami, either may be the first to arrive. If the first part to arrive at the shore is the ridge, a massive breaking wave or sudden flooding will be the first effect noticed on land. However, if the first part to arrive is a trough, a drawback will occur as the shoreline recedes dramatically, exposing normally submerged areas. The drawback can exceed hundreds of metres, and people unaware of the danger sometimes remain near the shore to satisfy their curiosity or to collect fish from the exposed seabed.

A typical wave period for a damaging tsunami is about twelve minutes. Thus, the sea recedes in the drawback phase, with areas well below sea level exposed after three minutes. For the next six minutes, the wave trough builds into a ridge which may flood the coast, and destruction ensues. During the next six minutes, the wave changes from a ridge to a trough, and the flood waters recede in a second drawback. Victims and debris may be swept into the ocean. The process repeats with succeeding waves.

Scales of Intensity and Magnitude

As with earthquakes, several attempts have been made to set up scales of tsunami intensity or magnitude to allow comparison between different events.

Intensity Scales

The first scales used routinely to measure the intensity of tsunamis were the *Sieberg-Ambraseys scale*, used in the Mediterranean Sea and the *Imamura-Iida intensity scale*, used in the Pacific Ocean. The latter scale was modified by Soloviev, who calculated the tsunami intensity "*I*" according to the formula:

$$I = \frac{1}{2} + \log_2 H_{av}$$

where H_{av} is the "tsunami height", averaged along the nearest coastline, with the tsunami height defined as the rise of the water level above the normal tidal level at the time of occurrence of the tsunami. This scale, known as the *Soloviev-Imamura tsunami intensity scale*, is used in the global tsunami catalogues compiled by the NGDC/NOAA and the Novosibirsk Tsunami Laboratory as the main parameter for the size of the tsunami.

This formula yields:

- $I = 2$ for $H_{av} = 2.8$ meters.
- $I = 3$ for $H_{av} = 5.5$ meters.
- $I = 4$ for $H_{av} = 11$ meters.
- $I = 5$ for $H_{av} = 22.5$ meters.
- etc.

In 2013, following the intensively studied tsunamis in 2004 and 2011, a new 12-point scale was proposed, the Integrated Tsunami Intensity Scale, intended to match as closely as possible to the modified ESI2007 and EMS earthquake intensity scales.

Magnitude Scales

The first scale that genuinely calculated a magnitude for a tsunami, rather than an intensity at a particular location was the ML scale proposed by Murty & Loomis based on the potential energy. Difficulties in calculating the potential energy of the tsunami mean that this scale is rarely used.

Abe introduced the *tsunami magnitude scale* M_t, calculated from,

$$M_t = a \log h + b \log R + D$$

where *h* is the maximum tsunami-wave amplitude (in m) measured by a tide gauge at a distance *R* from the epicentre, *a*, *b* and *D* are constants used to make the M_t scale match as closely as possible with the moment magnitude scale.

Tsunami Heights

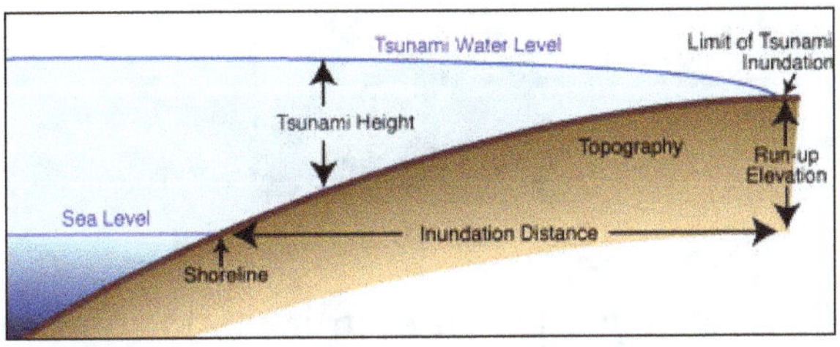

Diagram showing several measures to describe a tsunami size, including height, inundation and run-up.

Several terms are used to describe the different characteristics of tsunami in terms of their height:

- Amplitude, Wave Height, or Tsunami Height: Amplitude of Tsunami refers to its height relative to the normal sea level. It is usually measured at sea level, and it is different from the crest-to-trough height which is commonly used to measure other type of wave height.
- Run-up Height, or Inundation Height: The height reached by a tsunami on the ground above sea level, Maximum run-up height refers to the maximum height reached by water above sea level, which is sometimes reported as the maximum height reached by a tsunami.
- Flow Depth: Refers to the height of tsunami above ground, regardless of the height of the location or sea level.
- Maximum Water Level: Maximum height above sea level as seen from trace or water mark. Different from maximum run-up height in the sense that they are not necessarily water marks at inundation line/limit.

Warnings and Predictions

Drawbacks can serve as a brief warning. People who observe drawback (many survivors report an accompanying sucking sound), can survive only if they immediately run for high ground or seek the upper floors of nearby buildings. In 2004, ten-year-old Tilly Smith of Surrey, England, was on Maikhao beach in Phuket, Thailand with her parents and sister, and having learned about tsunamis recently in school, told her family that a tsunami might be imminent. Her parents warned others minutes before the wave arrived, saving dozens of lives. She credited her geography teacher, Andrew Kearney.

In the 2004 Indian Ocean tsunami drawback was not reported on the African coast or any other east-facing coasts that it reached. This was because the initial wave moved downwards on the eastern side of the megathrust and upwards on the western side. The western pulse hit coastal Africa and other western areas.

Tsunami warning sign.

A tsunami cannot be precisely predicted, even if the magnitude and location of an earthquake is known. Geologists, oceanographers, and seismologists analyse each earthquake and based on many factors may or may not issue a tsunami warning. However, there are some warning signs of an impending tsunami, and automated systems can provide warnings immediately after an earthquake in time to save lives. One of the most successful systems uses bottom pressure sensors, attached to buoys, which constantly monitor the pressure of the overlying water column.

Regions with a high tsunami risk typically use tsunami warning systems to warn the population before the wave reaches land. On the west coast of the United States, which is prone to Pacific Ocean tsunami, warning signs indicate evacuation routes. In Japan, the community is well-educated about earthquakes and tsunamis, and along the Japanese shorelines the tsunami warning signs are reminders of the natural hazards together with a network of warning sirens, typically at the top of the cliff of surroundings hills.

The Pacific Tsunami Warning System is based in Honolulu, Hawai'i. It monitors Pacific Ocean seismic activity. A sufficiently large earthquake magnitude and other information triggers a tsunami warning. While the subduction zones around the Pacific are seismically active, not all earthquakes generate a tsunami. Computers assist in analysing the tsunami risk of every earthquake that occurs in the Pacific Ocean and the adjoining land masses.

Tsunami hazard sign at Bamfield, British Columbia.

A tsunami warning sign in Kamakura, Japan.

A Tsunami hazard sign (Spanish - English) in Iquique, Chile.

Tsunami Evacuation Route signage along U.S. Route 101, in Washington.

As a direct result of the Indian Ocean tsunami, a re-appraisal of the tsunami threat for all coastal areas is being undertaken by national governments and the United Nations Disaster Mitigation Committee. A tsunami warning system is being installed in the Indian Ocean.

Computer models can predict tsunami arrival, usually within minutes of the arrival time. Bottom pressure sensors can relay information in real time. Based on these pressure readings and other seismic information and the seafloor's shape (bathymetry) and coastal topography, the models estimate the amplitude and surge height of the approaching tsunami. All Pacific Rim countries collaborate in the Tsunami Warning System and most regularly practise evacuation and other procedures. In Japan, such preparation is mandatory for government, local authorities, emergency services and the population.

One of the deep water buoys used in the DART tsunami warning system.

Along the United States west coast, in addition to sirens, warnings are sent on television and radio via the National Weather Service, using the Emergency Alert System.

Mitigation

In some tsunami-prone countries, earthquake engineering measures have been taken to reduce the damage caused onshore.

A seawall at Tsu, Japan.

Japan, where tsunami science and response measures first began following a disaster in 1896, has produced ever-more elaborate countermeasures and response plans. The country has built many tsunami walls of up to 12 metres (39 ft) high to protect populated coastal areas. Other localities have built floodgates of up to 15.5 metres (51 ft) high and channels to redirect the water from an incoming tsunami. However, their effectiveness has been questioned, as tsunamis often overtop the barriers.

The Fukushima Daiichi nuclear disaster was directly triggered by the 2011 Tōhoku earthquake and tsunami, when waves exceeded the height of the plant's sea wall. Iwate Prefecture, which is an area at high risk from tsunami, had tsunami barriers walls (Taro sea wall) totalling 25 kilometres (16 mi) long at coastal towns. The 2011 tsunami toppled more than 50% of the walls and caused catastrophic damage.

The Okushiri, Hokkaidō tsunami which struck Okushiri Island of Hokkaidō within two to five minutes of the earthquake on July 12, 1993, created waves as much as 30 metres (100 ft) tall—as high as a 10-storey building. The port town of Aonae was completely surrounded by a tsunami wall, but the waves washed right over the wall and destroyed all the wood-framed structures in the area. The wall may have succeeded in slowing down and moderating the height of the tsunami, but it did not prevent major destruction and loss of life.

Tornado

A tornado is a rapidly rotating column of air that is in contact with both the surface of the Earth and a cumulonimbus cloud or, in rare cases, the base of a cumulus cloud. The windstorm is often referred to as a twister, whirlwind or cyclone, although the word cyclone is used in meteorology to name a weather system with a low-pressure area in the center around which, from an observer looking down toward the surface of the earth, winds blow counterclockwise in the Northern Hemisphere and clockwise in the Southern. Tornadoes come in many shapes and sizes, and they are often visible in the form of a condensation funnel originating from the base of a cumulonimbus cloud, with a cloud of rotating debris and dust beneath it. Most tornadoes have wind speeds less than 110 miles per hour (180 km/h), are about 250 feet (80 m) across, and travel a few miles (several kilometers) before dissipating. The most extreme tornadoes can attain wind speeds of more than 300 miles per hour (480 km/h), are more than two miles (3 km) in diameter, and stay on the ground for dozens of miles (more than 100 km).

A tornado approaching Elie, Manitoba.

Various types of tornadoes include the multiple vortex tornado, landspout, and waterspout. Waterspouts are characterized by a spiraling funnel-shaped wind current, connecting to a large cumulus or cumulonimbus cloud. They are generally classified as non-supercellular tornadoes that develop over bodies of water, but there is disagreement over whether to classify them as true tornadoes. These spiraling columns of air frequently develop in tropical areas close to the equator and are less common at high latitudes. Other tornado-like phenomena that exist in nature include the gustnado, dust devil, fire whirl, and steam devil.

Tornadoes occur most frequently in North America (particularly in central and southeastern regions of the United States colloquially known as tornado alley), Southern Africa, northwestern and southeast Europe, western and southeastern Australia, New Zealand, Bangladesh and adjacent eastern India, and southeastern South America. Tornadoes can be detected before or as they occur through the use of Pulse-Doppler radar by recognizing patterns in velocity and reflectivity data, such as hook echoes or debris balls, as well as through the efforts of storm spotters.

Tornado Rating Scales

There are several scales for rating the strength of tornadoes. The Fujita scale rates tornadoes by damage caused and has been replaced in some countries by the updated Enhanced Fujita Scale. An F0 or EF0 tornado, the weakest category, damages trees, but not substantial structures. An F5 or EF5 tornado, the strongest category, rips buildings off their foundations and can deform large skyscrapers. The similar TORRO scale ranges from T0 for extremely weak tornadoes to T11 for the most powerful known tornadoes. Doppler radar data, photogrammetry, and ground swirl patterns (trochoidal marks) may also be analyzed to determine intensity and assign a rating.

Electromagnetic, Lightning and other Effects

Tornadoes emit on the electromagnetic spectrum, with sferics and E-field effects detected. There are observed correlations between tornadoes and patterns of lightning. Tornadic storms do not contain more lightning than other storms and some tornadic cells never produce lightning at all. More often than not, overall cloud-to-ground (CG) lightning activity decreases as a tornado touches the surface and returns to the baseline level when the tornado dissipates. In many cases, intense tornadoes and thunderstorms exhibit an increased and anomalous dominance of positive polarity CG discharges. Electromagnetics and lightning have little or nothing to do directly with what drives tornadoes (tornadoes are basically a thermodynamic phenomenon), although there are likely connections with the storm and environment affecting both phenomena.

Luminosity has been reported in the past and is probably due to misidentification of external light sources such as lightning, city lights, and power flashes from broken lines, as internal sources are now uncommonly reported and are not known to ever have been recorded. In addition to winds, tornadoes also exhibit changes in atmospheric variables such as temperature, moisture, and pressure. For example, on June 24, 2003 near Manchester, South Dakota, a probe measured a 100 mbar (hPa) (2.95 inHg) pressure decrease. The pressure dropped gradually as the vortex approached then dropped extremely rapidly to 850 mbar (hPa) (25.10 inHg) in the core of the violent tornado before rising rapidly as the vortex moved away, resulting in a V-shape pressure trace. Temperature tends to decrease and moisture content to increase in the immediate vicinity of a tornado.

Tornado Preparedness

The term "tornado preparedness" refers to safety precautions made before the arrival of and during a tornado. Historically, the steps taken have varied greatly, depending on location, or time remaining before a tornado was expected. For example, in rural areas, people might prepare to enter an external storm cellar, in case the main building collapses, and thereby allow exit without needing rescue from the main building as in urban areas. Because tropical storms have spawned many tornadoes, hurricane preparations also involve tornadoes. The term "tornado preparedness" has been used by government agencies, emergency response groups, schools, insurance companies, and others.

Understanding the Dangers

Preparedness involves knowing the major dangers to avoid. Some tornadoes are the most violent storms in nature. Tornadoes have varied in strength, and some tornadoes have been mostly invisible due to a lack of loose dirt or debris in the funnel cloud. Spawned from severe thunderstorms, tornadoes have caused fatalities and devastated neighborhoods within seconds of arrival.

A tornado with no obvious funnel from the upper clouds, although the rotating dust cloud indicates strong winds at the surface.

A tornado operates as a rotating, funnel-shaped cloud that extends downward from a thunderstorm, to the ground, with swirling winds which have reached 300 miles per hour (480 km/h). The wind speed might be difficult to imagine: traveling the length of a U.S. football field within 1 second (over 130 meters or 430 feet per second). Damage paths have been in excess of one-mile wide (1.6 km) and 50 miles long (80 km).

Not all tornadoes are easily seen. A tornado funnel can be transparent until reaching an area with loose dirt and debris. Also, some tornadoes have been seen against sunlit areas, but rain or nearby low-hanging clouds has obscured other tornadoes. Occasionally, tornadoes have developed so suddenly, so rapidly, that little, if any, advance warning was possible.

Before a tornado strikes an area, the wind has been known to die down and the air to become very still. A cloud of debris has sometimes marked the bottom of a tornado even when the funnel was not visible. Tornadoes typically occur along the trailing edge of a thunderstorm.

The following is a summary of typical tornadoes:

- They may strike quickly, with little or no warning.
- They may appear nearly transparent until dust and debris are picked up or a cloud forms in the funnel.
- The average tornado moves Southwest to Northeast in the U.S., but tornadoes have been known to move in any direction.
- The average forward speed of a tornado is 30 miles per hour (48 km/h), but has varied from stationary to 70 mph (110 km/h).

- Tornadoes can also accompany tropical storms and hurricanes as they move onto land.
- Waterspouts are tornadoes that form over water.
- Tornadoes are most frequently reported east of the Rocky Mountains during spring and summer months.
- Peak tornado season in the southern U.S. states is March through May; in the northern states, it is late spring through early summer.
- Tornadoes are most likely to occur between 3 p.m. and 9 p.m. (local time), but have occurred at other times.

Steps when Expecting Storms to Arrive

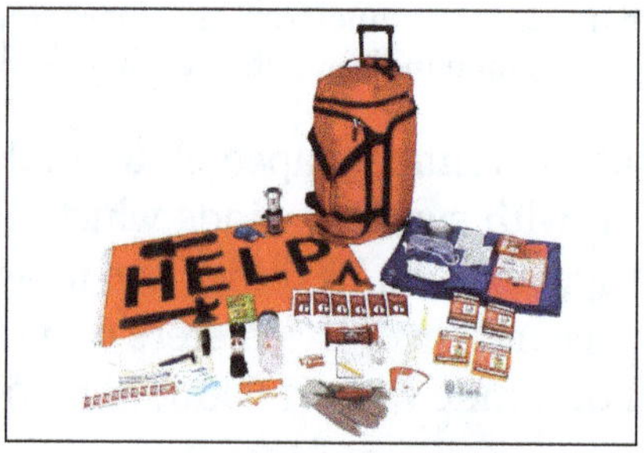

Example of a tornado emergency kit.

The U.S. Federal Emergency Management Agency (FEMA) has advised the following precautions before a storm reaches an area:

- Be alert to the changing weather conditions.
- Listen to NOAA Weather Radio and/or Skywarn, or to local commercial radio or television newscasts for the latest information.
- Watch various common danger signs, including:
  - Large hail stones.
  - A large, dark, low-lying cloud (particularly if rotating).
  - Loud roar of wind, sounding similar to a freight train.

Upon seeing an approaching storm or noticing any of the danger signs, they were advised to prepare to take shelter immediately, such as moving to a safe room, internal stairway, or other safe-haven area.

All individuals and families should have a disaster preparedness kit made prior to tornado. According to FEMA the kit should include items needed to shelter in place in the event of a disaster such as a tornado for up to 72 hours following impact.

Actions taken during Tornadoes

Table: During August 2010, FEMA advised people to perform the following actions when a tornado struck.

Location	Action taken
In a structure (e.g. residence, small building, school, nursing home, hospital, factory, shopping center, high-rise building,restaurant).	They were to enter a pre-designated shelter area such as a safe room, basement, storm cellar, or the lowest building level. If there was no basement, then to the center of an interior room on the lowest level (closet, interior hallway) away from corners, windows, doors, and outside walls. The goal has been to put as many walls as possible between there and the outside. They were advised to get under a sturdy table and use arms to protect head and neck, and not open windows.
In a vehicle, trailer, or mobile home	They were advised to leave immediately and enter the lowest floor of a sturdy, nearby building or a storm shelter. Mobile homes, even if tied down, offer little protection from tornadoes. If a car is flipped by high winds, there is also the danger of broken glass.
On the outside with no shelter	They were advised to lie flat in a nearby ditch or depression and cover head with their hands. Also, to beware of the potential for flooding there.
	They were advised to not stay under an overpass or bridge (where winds or debris might be funneled). It was safer to be in a low, flat location.
	The advice was to never try to outrun a tornado in urban or congested areas in a car or truck, but instead, to leave the vehicle immediately for safe shelter.
	Flying debris from tornadoes causes most fatalities and injuries.

Because some preparations vary, depending on location, people have been advised to consult their local area preparedness plans, rather than assume the plans are similar for all areas, such as which local buildings have been designated as storm shelters.

A 2012 study of tornado injuries found that wearing a helmet such as those used for American football or bicycling, is an effective way to reduce injuries and deaths from head trauma. As of 2012, the CDC endorsed only general head protection, but recommended that if helmets are to be used, they be kept close by to avoid wasting time searching for them.

After the 2013 Moore tornado, it became apparent that thousands of people attempt to flee major tornadoes, and this has been credited with reducing the death toll. However, during this event some people were killed as the tornado passed over the traffic jam caused by the impromptu evacuation. In addition to urban traffic, evacuation can also be hampered by flash flooding produced by associated thunderstorms, and the need to be certain about the position and direction of the tornado. Others who did not flee the

Moore tornado were also killed because the buildings they were hiding in were completely destroyed, highlighting the need for storm shelters and safe rooms constructed specifically to withstand very high winds.

Long-term Preparations

Depending on location, various safe-haven areas have been prepared. The goal has been to avoid outer walls which might collapse when a roof section becomes airborne and the walls below lose their upper support: many interior rooms resist collapse longer, due to smaller walls interconnected to each other, while outer walls deflect the force of the winds. Because mobile homes typically lack foundation anchors and present a large surface-area sail (to catch wind), the advice has been to seek a safe haven elsewhere, such as in a stronger nearby building. When a mobile home begins to roll, people have been injured by hitting objects inside, or being crushed when a trailer suddenly hits the ground and begins to collapse around them.

In a multi-story building, an internal stairway (away from broken windows) often acts as a safe haven, due to the stairs reinforcing the walls and blocking any major debris falling from above. If a stairway is lined with windows, then there would be the danger of flying glass, so an interior stairway, or small inner room, would be preferable.

In private homes, some similar stairway rooms have been used, or an interior room/ closet kept clear to quickly allow entry when a storm is seen or heard approaching (the wind roar intensifies, sounding like a swift "freight train" coming nearer, louder). With weeks or months to prepare, an interior safe room can be constructed, with space for emergency water, food and flashlights, and a telephone to call for rescue if the exit becomes blocked by falling debris. Some above-ground safe rooms have been built with steel-rebar rods in cement-filled cinder blocks, to withstand winds of 250 miles per hour (400 km/h). Rural homes might have an outside storm cellar, or other external bunker, to avoid being trapped within a collapsing house. In rural homes, generators are also helpful to maintain power with enough fuel for a few days.

There were no building codes requiring tornado shelters nor specifically designed to prevent tornado damage until the 2011 Joplin tornado prompted a local ordinance requiring hurricane ties or similar fasteners. The state of Oklahoma adopted the minimum U.S. standard that year for the first time, but did not add high-wind protections like those in Florida designed to protect against hurricanes. Other states in Tornado Alley have no statewide building codes. The chance of any given location in Tornado Alley getting hit by an F-2 tornado (strong enough to do major structural damage and exceeding the 90 mph guideline for straightline winds) is about 1 every 4,000-5,000 years; in other areas the annual probability is one in several million. The most stringent building codes only require earthquake strengthening for a 1 in every 500-1,000 year probability.

The U.S. Federal Emergency Management Agency has spent tens millions of dollars subsidizing the construction of shelters and safe rooms in both private and public buildings. Many buildings in Tornado Alley do not have basements, because unlike in more northern areas, there is no need for a deep foundation to get below the frost line, in some places the water table is high, and expansion and contraction of clay-heavy soils can produce additional pressure on buildings that can cause leaks if not reinforced.

Medical Preparations

Having a first aid kit in the safe haven is advised to help victims recover from minor injuries. People needing prescription medications could have a medicine bag ready to take to shelter. Some people have reported their "ears popping" due to the change in air pressure, but those effects seem to be temporary. Covering people with mattresses or cushions has helped avoid injury from flying debris, as walls collapsed nearby.

Injuries sustained during a tornado vary in nature and in severity. The most common injuries experienced during a tornado are complex contaminated soft tissue wounds and account for more than 50% of the cases seen by emergency rooms following a tornado. These wounds will most likely be contaminated with soil and foreign bodies due to high wind speeds caused by tornadoes. Fractures are the second most common injury obtained after a tornado strikes and account for up to 30% of total injuries. Head injuries are also commonly reported during a tornado, but severe head injuries only account for less than 10% of the total. Even though only 10% of reported head injuries are severe, they are the most common cause of death following a tornado. Blunt trauma to the chest and abdomen are also injuries obtained following a tornado, but only account for less than 10% of overall injuries.

Tornado Drills

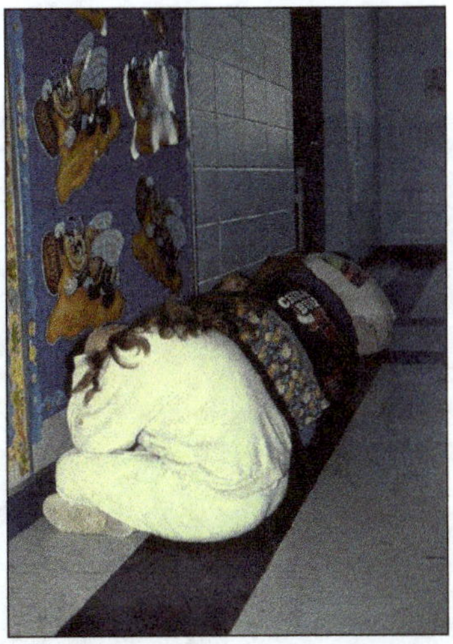

Students participate in a tornado drill, lining up along an interior wall and covering their heads.

Tornado drills are an important element in tornado preparedness. Like any other safety drills, they increase chances of correct response to a real tornado threat.

Most states in the midwestern and southern United States conduct a statewide tornado drill in late winter or early spring in preparation for the severe weather season. During these drills, the National Weather Service issues test tornado warnings, and local Emergency Alert Systems and/or NOAA Weather Radio (normally as a Required Weekly Test or Required Monthly Test; Live Tornado Warning Codes can only be used if a waiver from the FCC is granted since "Live Code Testing" is prohibited per regulations) are activated along with outdoor warning sirens. Schools and businesses may also conduct a tornado drill simultaneously.

A tornado drill is a procedure of practicing to take cover in a specified location in the event that a tornado strikes an area. This safety drill is an important element of tornado preparedness.

Generally, a signal is given, such as a series of tones (ex. Continuous Tone), or a voice announcement. Upon receiving the signal, building occupants of schools, hospitals, factories, shopping centers, etc. proceed to a designated location, usually an interior room or corridor with no windows, and assume a protective position.

In homes and small buildings one must go to the basement or an interior room on the lowest floor (closet, bathroom), to stay away from glass.

Cars and mobile homes must be abandoned, if there is small chance to drive away.

In some jurisdictions, schools are required to conduct regular tornado drills, though generally less frequently than fire drills.

Landslide

A landslide near Cusco, Peru in 2018.

The term landslide or less frequently, landslip, refers to several forms of mass wasting that include a wide range of ground movements, such as rockfalls, deep-seated slope failures, mudflows, and debris flows. Landslides occur in a variety of environments, characterized by either steep or gentle slope gradients, from mountain ranges to coastal cliffs or even underwater, in which case they are called submarine landslides. Gravity is the primary driving force for a landslide to occur, but there are other factors affecting slope stability that produce specific conditions that make a slope prone to failure. In many cases, the landslide is triggered by a specific event (such as a heavy rainfall, an earthquake, a slope cut to build a road, and many others), although this is not always identifiable.

Causes

The Mameyes Landslide, in the Mameyes neighborhood of barrio Portugués Urbano in Ponce, Puerto Rico, which buried more than 100 homes, was caused by extensive accumulation of rains and, according to some sources, lightning.

Landslides occur when the slope (or a portion of it) undergoes some processes that change its condition from stable to unstable. This is essentially due to a decrease in the shear strength of the slope material, to an increase in the shear stress borne by the material, or to a combination of the two. A change in the stability of a slope can be caused by a number of factors, acting together or alone. Natural causes of landslides include:

- Saturation by rain water infiltration, snow melting, or glaciers melting.
- Rising of groundwater or increase of pore water pressure (e.g. due to aquifer recharge in rainy seasons, or by rain water infiltration).
- Increase of hydrostatic pressure in cracks and fractures.
- Loss or absence of vertical vegetative structure, soil nutrients, and soil structure (e.g. after a wildfire – a fire in forests lasting for 3–4 days).
- Erosion of the toe of a slope by rivers or ocean waves.
- Physical and chemical weathering (e.g. by repeated freezing and thawing, heating and cooling, salt leaking in the groundwater or mineral dissolution).

- Ground shaking caused by earthquakes, which can destabilize the slope directly (e.g., by inducing soil liquefaction) or weaken the material and cause cracks that will eventually produce a landslide.
- Volcanic eruptions.

Landslides are aggravated by human activities, such as:

- Deforestation, cultivation and construction.
- Vibrations from machinery or traffic.
- Blasting and mining.
- earthwork (e.g. by altering the shape of a slope, or imposing new loads).
- In shallow soils, the removal of deep-rooted vegetation that binds colluvium to bedrock.
- Agricultural or forestry activities (logging), and urbanization, which change the amount of water infiltrating the soil.

Types

Debris Flow

Slope material that becomes saturated with water may develop into a debris flow or mud flow. The resulting slurry of rock and mud may pick up trees, houses and cars, thus blocking bridges and tributaries causing flooding along its path.

Debris flow is often mistaken for flash flood, but they are entirely different processes.

Muddy-debris flows in alpine areas cause severe damage to structures and infrastructure and often claim human lives. Muddy-debris flows can start as a result of slope-related factors and shallow landslides can dam stream beds, resulting in temporary water blockage. As the impoundments fail, a "domino effect" may be created, with a remarkable growth in the volume of the flowing mass, which takes up the debris in the stream channel. The solid–liquid mixture can reach densities of up to 2,000 kg/m^3 (120 lb/cu ft) and velocities of up to 14 m/s (46 ft/s). These processes normally cause the first severe road interruptions, due not only to deposits accumulated on the road (from several cubic metres to hundreds of cubic metres), but in some cases to the complete removal of bridges or roadways or railways crossing the stream channel. Damage usually derives from a common underestimation of mud-debris flows: in the alpine valleys, for example, bridges are frequently destroyed by the impact force of the flow because their span is usually calculated only for a water discharge. For a small basin in the Italian Alps (area 1.76 km^2 (0.68 sq mi)) affected by a debris flow, estimated a peak discharge of 750 m^3/s (26,000 cu ft/s) for a section located in the middle stretch of the main channel. At the same cross section, the maximum foreseeable water discharge (by HEC-1), was 19 m^3/s (670 cu ft/s), a value about 40 times lower than that calculated for the debris flow that occurred.

Earthflow

The Costa della Gaveta earthflow in Potenza, Italy. Even though it moves just some mm/a and is hardly visible, this landslide causes progressive damage to the national road, the national highway, a flyover and several houses that were built on it.

An earthflow is the downslope movement of mostly fine-grained material. Earthflows can move at speeds within a very wide range, from as low as 1 mm/yr (0.039 in/yr) to 20 km/h (12.4 mph). Though these are a lot like mudflows, overall they are more slow moving and are covered with solid material carried along by flow from within. They are different from fluid flows which are more rapid. Clay, fine sand and silt, and fine-grained, pyroclastic material are all susceptible to earthflows. The velocity of the earthflow is all dependent on how much water content is in the flow itself: the higher the water content in the flow, the higher the velocity will be.

A rock slide in Guerrero, Mexico.

These flows usually begin when the pore pressures in a fine-grained mass increase until enough of the weight of the material is supported by pore water to significantly decrease the internal shearing strength of the material. This thereby creates a bulging lobe which advances with a slow, rolling motion. As these lobes spread out, drainage of the mass increases and the margins dry out, thereby lowering the overall velocity of the flow. This process causes the flow to thicken. The bulbous variety of earthflows are not that spectacular, but they are much more common than their rapid counterparts. They develop a sag at their heads and are usually derived from the slumping at the source.

Earthflows occur much more during periods of high precipitation, which saturates the ground and adds water to the slope content. Fissures develop during the movement of clay-like material which creates the intrusion of water into the earthflows. Water then increases the pore-water pressure and reduces the shearing strength of the material.

Debris Slide

A debris slide is a type of slide characterized by the chaotic movement of rocks, soil, and debris mixed with water and/or ice. They are usually triggered by the saturation of thickly vegetated slopes which results in an incoherent mixture of broken timber, smaller vegetation and other debris. Debris avalanches differ from debris slides because their movement is much more rapid. This is usually a result of lower cohesion or higher water content and commonly steeper slopes.

Steep coastal cliffs can be caused by catastrophic debris avalanches. These have been common on the submerged flanks of ocean island volcanos such as the Hawaiian Islands and the Cape Verde Islands. Another slip of this type was Storegga landslide.

Goodell Creek Debris Avalanche, Washington, USA.

Debris slides generally start with big rocks that start at the top of the slide and begin to break apart as they slide towards the bottom. This is much slower than a debris avalanche. Debris avalanches are very fast and the entire mass seems to liquefy as it slides down the slope. This is caused by a combination of saturated material, and steep slopes. As the debris moves down the slope it generally follows stream channels leaving a v-shaped scar as it moves down the hill. This differs from the more U-shaped scar of a slump. Debris avalanches can also travel well past the foot of the slope due to their tremendous speed.

Rock Avalanche

A rock avalanche, sometimes referred to as sturzstrom, is a type of large and fast-moving landslide. It is rarer than other types of landslides and therefore poorly understood. It exhibits typically a long run-out, flowing very far over a low angle, flat, or even slightly uphill terrain. The mechanisms favoring the long runout can be different, but they typically result in the weakening of the sliding mass as the speed increases.

Blockade of Hunza river.

Shallow Landslide

A landslide in which the sliding surface is located within the soil mantle or weathered bedrock (typically to a depth from few decimeters to some meters) is called a shallow landslide. They usually include debris slides, debris flow, and failures of road cut-slopes. Landslides occurring as single large blocks of rock moving slowly down slope are sometimes called block glides.

Hotel Panorama at Lake Garda. Part of a hill of Devonian shale was removed to make the road, forming a dip-slope. The upper block detached along a bedding plane and is sliding down the hill, forming a jumbled pile of rock at the toe of the slide.

Shallow landslides can often happen in areas that have slopes with high permeable soils on top of low permeable bottom soils. The low permeable, bottom soils trap the water in the shallower, high permeable soils creating high water pressure in the top soils. As the top soils are filled with water and become heavy, slopes can become very unstable and slide over the low permeable bottom soils. Say there is a slope with silt and sand as its top soil and bedrock as its bottom soil. During an intense rainstorm, the bedrock will keep the rain trapped in the top soils of silt and sand. As the topsoil becomes saturated and heavy, it can start to slide over the bedrock and become a shallow landslide. R. H. Campbell did a study on shallow landslides on Santa Cruz Island, California. He notes that if permeability decreases with depth, a perched water table may develop in soils at intense precipitation. When pore water pressures are sufficient to reduce effective normal stress to a critical level, failure occurs.

Deep-seated Landslide

Deep-seated landslides are those in which the sliding surface is mostly deeply located below the maximum rooting depth of trees (typically to depths greater than ten meters). They usually involve deep regolith, weathered rock, and/or bedrock and include large slope failure associated with translational, rotational, or complex movement. This type of landslide potentially occurs in an tectonic active region like Zagros Mountain in Iran. These typically move slowly, only several meters per year, but occasionally move faster. They tend to be larger than shallow landslides and form along a plane of weakness such as a fault or bedding plane. They can be visually identified by concave scarps at the top and steep areas at the toe.

Deep-seated landslide on a mountain in Sehara, Kihō, Japan caused by torrential rain of Tropical Storm Talas.

Landslide Preparedness Checklist

- Be conscious of landslide and debris-flow risk when buying a home or property.
- Make a disaster supply kit and have a family emergency plan.

- Follow proper land-use procedures: avoid building near steep slopes, close to mountain edges, near drainage ways or along natural erosion valleys.
- Become familiar with the land around you; learn whether debris flows have occurred in your area by contacting local officials — slopes where debris flows have occurred in the past are likely to experience them in the future.
- Get a ground assessment of your property.
- Consult a professional for advice on appropriate preventative measures for your home or business, such as flexible pipe fittings, which can better resist breakage.
- Protect your property by planting ground cover on slopes and building retaining walls.
- In mudflow areas, build channels or deflection walls to direct the flow around buildings, but be aware, if you build walls to divert debris flow and the flow lands on a neighbor's property, you may be liable for damages.
- If you are at risk from a landslide talk to your insurance agent — debris flow may be covered by flood insurance policies from the National Flood Insurance Program (NFIP).

During Heavy Rainfall or a Landslide:

- During heavy rain, stay alert and awake; many deaths from landslides occur while people are sleeping.
- Listen to NOAA Weather Radio and local news stations on a battery-powered radio for warnings of heavy rainfall.
- Listen for unusual sounds that might indicate moving debris, such as trees cracking or boulders knocking together.
- Move away from the path of a landslide or debris flow as quickly as possible. The danger from a mudflow increases near stream channels and with prolonged heavy rains. Mudflows can move faster than you can walk or run. Look upstream before crossing a bridge and do not cross the bridge if a mudflow is approaching.
- Avoid river valleys and low-lying areas.
- If you are near a stream or channel, be alert for any sudden increase or decrease in water flow and notice whether the water changes from clear to muddy. Such changes may mean there is debris flow activity upstream so be prepared to move quickly.
- Curl into a tight ball and protect your head if escape is not possible.

References

- Michael H Mogil (2007). Extreme Weather. New York: Black Dog & Leventhal Publisher. pp. 210–11. ISBN 978-1-57912-743-5
- Earthquake-geology, science: britannica.com, Retrieved 17 August, 2020
- A. J. Bedard (January 2005). "Low-Frequency Atmospheric Acoustic Energy Associated with Vortices Produced by Thunderstorms". Mon. Wea. Rev. 133 (1): 241–63. Bibcode:2005MWRv..133..241B. doi:10.1175/MWR-2851.1
- Landslide, prepare: wunderground.com, Retrieved 08 June, 2020
- "Ancient Volcano Collapse Caused A Tsunami With An 800-Foot Wave". Popular Science. Retrieved 2017-10-20

3
Preparedness and Response

Preparedness involves research, estimation and planning of precautionary measures to counter the potential disasters. Disaster response includes searching and rescuing, providing immediate assistance and restoration of infrastructure for supporting life of the affected population. All the aspects related to preparedness and response in emergency management have been carefully analyzed in this chapter.

Preparedness

The core element of the effective disaster mitigation is preparedness. The systematic process of using administrative measures, organizations, and operational skills to implement strategies, policies, and improved coping capacities in order to lessen the adverse impacts of hazards and to minimize the opportunity for development of emergencies (disasters) is called emergency (disaster) preparedness.

Preparedness means proper planning, resource allocation, and training, including simulated disaster response exercises. Preparedness is always based on a sound risk assessment analysis and good linkages with early warning systems. The risk assessment analysis is composed of two types of risk assessment:

- Early risk assessment.
- Initial risk assessment.

Three general priorities should be identified for early risk assessment: locating the problem, determining the magnitude of the problem, and determining of immediate priorities. The initial needs assessment is based on the situation analyses of available resources and there are about 5 essential (set as minimum) needs during the disaster: water and sanitation, health services, shelter, food and nutrition and coordination activities.

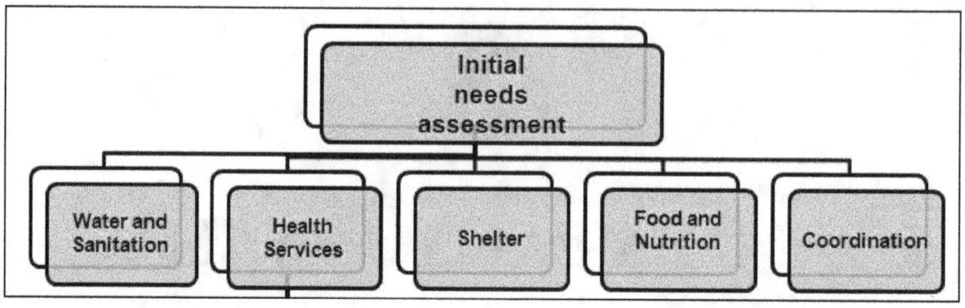

Preparedness means developing a proper plan for ensuring that all elements of the early and initial risk assessments are well recognized and mechanisms for minimizing the risk and saving of lives and properties are appropriately established and coordinated. The main goal of preparedness is to create a readiness which is described as an ability to quickly and appropriately respond when required. Different emergencies/disasters result in different consequences. Therefore, the preparedness process would be created as a generic preparedness i.e. "all-hazard preparedness" or "specific".

hazard's related preparedness process (preparedness for natural disasters, biological and chemical attacks and accidents, nuclear and radiological preparedness, terrorist attacks etc.). The U.S. Homeland Security Department has developed 15 different preparedness and planning scenarios for different types of hazards.

Additionally, specific preparedness plan may be targeted for the protection of the critical infrastructure of the community, aiming to reduce vulnerabilities and strengthen the resilience of critical infrastructure across the ten sectors:

• Health	• Safety
• Food	• Energy and utilities
• Finance	• Manufacturing
• Water	• Government
• Information and Communication Technology	• Transportation

Critical infrastructure is the backbone of modern society and it is essential to national prosperity. The infrastructure enables a nation's productivity, quality of life, and economic progression by driving growth, creating jobs and improving productivity, quality of life and efficiency. Networked infrastructure, such as transportation and communications, boosts long-term economic outputs.

Responsibility and End-users of the Preparedness Planning

Preparedness must be supported by formal institutions and to be based on their legal and budgetary capacities. Preparedness actions and activities can be divided according to the responsibility for implementation and possible recipients/users. Therefore, preparedness activities have two components: governmental and population based preparedness. The government component, which includes the obligation and responsibility for disaster/emergency management, including overall preparedness of governmental and other service agencies and activities, is one group. Population/individuals and businesses preparedness activities are the second component of the complex process entitled as disaster/emergency preparedness. All activities covered by this group are part of the governmental preparedness planning activities, but they are directly targeted to the end users - citizens and business community.

Governmental Preparedness

Almost all governments maintain a relatively universal set of systems and tools that address the risk posed by natural, technological, and intentional hazards. Although each nation's disaster management components and systems have developed independently and are rooted in different histories, economic realities and other factors, vast institutional sharing between governments has led to a growing organizational standardization of practices (SOPs), protocols and equipment used by disaster organizations. Most of those good practices are notably in the area of first response: fire departments, law enforcement agencies (police departments), emergency management (civil protection) agencies, emergency medical services, and the military. Although certain factors such as wealth, technical expertise, govern in accordance with the specificity of the each country, governmental fundamental missions are almost identical and their preparedness actions should be developed following the common preparedness strategy.

The diverse range of government preparedness actions may be grouped into five general categories, even though all of them must be covered by general planning activities:

- Planning.
- Resources and equipment.
- Exercise.
- Training.
- Statutory authority.

Planning: The most comprehensive methodology used by the governments at every level (from local to national) to plan for disasters is the creation of a community or national Emergency Operation Plan (EOP). These plans can be scaled up or down depending on event-specific disaster response and recovery requirements.

An EOP is a document that describes in intricate detail the people and agencies who will be involved in the response to the emergencies/disasters and their responsibilities. It also describes how citizens and structures will be protected in the event of a disaster. It may catalog the equipment, facilities, and resources available within and outside the jurisdiction. EOPs are also referred to as contingency plans, continuity of operations plans, emergency response plans, and counter-disaster plans. They can also be created for individual entities, such as schools, hospitals, prisons, and utilities facilities. The main purpose is to introduce and describe various concepts and policies, clarify individual and agency responsibilities, and delineate authority.

According to the preparedness strategy, EOPs should be develop as a generic preparedness plans ("all hazards") or as a hazard's specific preparedness plan; single agency or multi-agency; local, regional or national and as business continuity plans.

All-hazards plan approach is based on the premise of an organization's (service's) response to the range of all potential major incidents. Single generic plan can provide a basic structured response for any incident including chemicals, fuel, electricity, flooding etc.

Specific plans approach is designed to meet specific needs (risk specific, site specific or organization's function specific). Multi-agency/integrated emergency plan aims to ensure that the activities of all services/ organizations involved in managing a major incident operate in an integrated manner.

Resources and equipment: The set of tools, technology, and other equipment to assist in disaster response and recovery has helped response agencies to drastically reduce the numbers of injuries and deaths and the amount of property damaged and destroyed as a result of disaster events. This equipment has also increased the effectiveness of response agencies by protecting the lives of the responders. Unfortunately, access to this equipment is driven primarily by the resources available, so there exists great disparity throughout the world in terms of who has what equipment. The equipment means to buy or to mobilize in case of disasters, vehicles, communication equipment, equipment for medical care, personal protective equipment etc.

Exercise: A major part of the preparedness in pursuit of response capability is a disaster exercise regimen. Response exercises allow those involved in emergency and disaster response, as defined in the EOP, to practice and validate their roles and responsibilities before an actual event occurs. Exercises should be designed to engage team members and get them working together to manage the response to a hypothetical incident. Exercises enhance knowledge of plans, allow members to improve their own performance and also help discover any unforeseen problems in the plan in a non-emergency situation. Five kinds of exercises can be conducted in the name of emergency preparedness:

- Walkthroughs, workshops or orientation seminars.
- Tabletop exercise.
- Drill.
- Functional exercise.
- Full-scale exercise.

Walkthroughs, workshops and orientation seminars are basic training for team members. They are designed to familiarize team members with emergency response, business continuity and crisis communications plans and their roles and responsibilities as defined in the plans.

Tabletop exercises are discussion-based sessions where team members meet in an informal, classroom setting to discuss their roles during an emergency and their responses to a particular emergency situation. A facilitator guides participants through a discussion of one or more scenarios.

A drill is a controlled, supervised method by which a single disaster management operation or function is practiced or tested. Drills are most effective when they mimic real-life situations. The tabletop exercise is designed to allow disaster management officials to practice the full activation of the emergency response plan within the confines of a controlled, low-stress discussion scenario, leaded by facilitator.

During the functional exercise are tested and practiced the capabilities of disaster managers by simulating (in time depending manner) an event to which they must respond. The full-scale exercise is a scenario-based event that seeks to create an atmosphere closely mimicking an actual disaster. All players are required to act during a real event, as outlined in the EOP, working in real time and using all of the required equipment and procedures.

Training: Training is the fourth component to government preparedness. It goes without saying that disaster response officials are more effective if they are trained to do their jobs. Disaster management and disaster preparedness training is not universally available. Even though first-response officials, namely police, fire, and emergency medical services (EMS), are likely to have some basic standard of introductory training no matter where they are located, the specialized instruction required for disaster response is much more technical. Several nations have established centralized or regional training facilities/centers that provide these skills. Some of them work on development and improvement of the teaching capabilities using different approach. The most important specialized training modules that aim to improve the skills primarily of fire, police, and EMS officials, include evacuation, mass care, mass fatalities management, debris management, flood-fighting operations, warning coordination, hazardous materials, weapons of mass destruction, radiological response, crowd control, response to terrorist attacks, wildfire and wild land fire response.

Statutory authority: The final link in government emergency preparedness is the statutory authority. Government response actions involve a diverse range of government officials and agencies interacting with the public and businesses, and operating on public and private land. There are often broad expenditures of funds, suspensions of normal government and private activities, and other major deviations from–normal. To ensure that all individuals and agencies involved in the emergency management system are able to carry out their duties, it is vitally important that the proper statutory authorities exist. Statutory authorities ensure that emergency and disaster response agencies and functions are established, staffed, and receive regular funding. Largely as a result of UN efforts, namely the Hyogo Framework for Action, almost all countries have, at least in concept, created an office at the national government level that manages emergency situations. The national government emergency management agencies are the best places to provide emergency management priorities, standards, direction, and goals to guide local emergency managers; provide training and expertise in the field of emergency management; provide funding to support mitigation, preparedness, response, and recovery; provide specialized assets, which could include hazardous materials detection, containment, cleanup, and decontamination; heavy lifting and debris removal equipment; and infrastructure repair teams and equipment.

Public Preparedness

Public preparedness (individuals and business community) can be considered as actions taken to empower ordinary citizens to help themselves, their families, their neighbors, and complete strangers. To be effective, this effort must exceed simply raising awareness of a hazard. A prepared public must be given the skills that allow them to perform specialized actions such as search and rescue, first aid, and fire suppression. Individual and home preparedness begins with a written plan. The plan needs to consider a variety of scenarios. Consider what you would need to do to sustain yourself at home for several days without utilities or communication. If your home is damaged, you need to have a neighborhood meeting location identified nearby where everyone in the family can meet. If you are at work, school, or someplace else, you need to identify a regional location where you will meet.

The U.S. Department of Health and Senior Services has developed a program called "Ready in 3". This program encourages individuals to take three simple steps to become better prepared: create a plan, prepare a kit, and listen to information. Recommendations for an emergency kit include support for every family member and pets for at least 3 days. This is an excellent baseline level of preparedness and it is kept simple and inexpensive to allow those short on time or resources to make a step in the right direction.

Public education for improvement of the public preparedness, also called risk communication, preparedness education, social marketing or disaster education, is the backbone of any effective public preparedness effort. Public education efforts have three main goals: awareness of the hazard risk, behavior change, and warning.

The possible methods by which disaster managers may educate the public are numerous and diverse: mass media communication, advertisements, posters, endorsements, employee preparedness campaigns, schools, business, churches, libraries, special events, one-to-one meetings, social networks etc. Feasibility and audience suitability are the key factors in choosing the appropriate method.

Public Health Emergencies

Nowadays, all public health experts must be aware that ideas, information, and microbes are shared worldwide more easily than ever before. New infections, such as the novel influenza A, H7N9, Middle East respiratory syndrome corona virus or Zika-virus diseases, pay little heed to political boundaries as they spread. The number of terrorist's attacks is bigger and bigger from year to year and the humanitarian emergencies are pressing very strong on the national borders, giving them the high potential to influence the public health of all affected countries, creating a new cross border public health emergency. It's important to stress that emergencies can come in ever-complicated combination: a combined earthquake, tsunami, and radiation disaster; terrorist attacks and bombings that require as much psychological profiling as chemical profiling; or climate change and environmental pollution that would change the pattern of communicable diseases.

Public health emergencies are multidimensional, dynamic situations that overwhelm existing healthcare and public health infrastructure resulting in adverse community health effects.

The ensuring of the security of a population's health entails preventing, protecting, mitigating, responding to, and recovering from a wide range of hazards and possible health related consequences. In an emergency, the capacities of all sectors are used to mitigate the acute event. However, the public health consequences of an event are not always visible, and, historically, health expertise has been conspicuously absent from emergency management.

Over the past decade, there is a growing public awareness that health is part of almost every event. In a case of emergency, the scope of public health activities is extends from local to national levels. Moreover, global public health responses involve cross-border health risks, including pandemics, climate change, famine, and displaced populations. Much progress has been made in emergency management to use public health expertise in all phases of the cycle: planning, response, and recovery.

Those complex and interconnected problems have spurred innovation to create interconnected solutions. Increasingly, the countries around the world started to build their national capabilities aiming to improve the health security. Some nations and local communities have well-developed public health emergency plans. But, the fact that most communities are without written and coordinated preparedness and response plans, which offer their citizens the most effective and comprehensive protection in case of public health emergencies, created new approaches headed by international organizations and national governments.

Public Health Emergency Preparedness

The preparedness for and the response to all health emergencies require up-to-date knowledge and skills and prepared workforce to cope with the challenge to protect the health of citizens. However, sometimes it requires very sophisticated equipment such as laboratories with genetic sequencing capacities for infectious agents and rapid detection methods for chemical and radiological hazards. The presence of nimble medical and epidemiological response units, vigilant surveillance, and early warning systems are essential for appropriate and effective public health preparedness.

The public health preparedness cycle is the same as well the disaster preparedness cycle. The main goal of effective public health emergency preparedness is to create public health system, which will be able to respond effectively, aiming to protect the health of the population and to sustain the health security.

Public health emergency preparedness activities are all activities taken by healthcare and public health organizations to ensure effective response to emergencies that impact health, especially events that have timing or scale that overwhelms normal capacity. Therefore, public health preparedness must be developed following all specificity of the public health and in accordance with emergency preparedness.

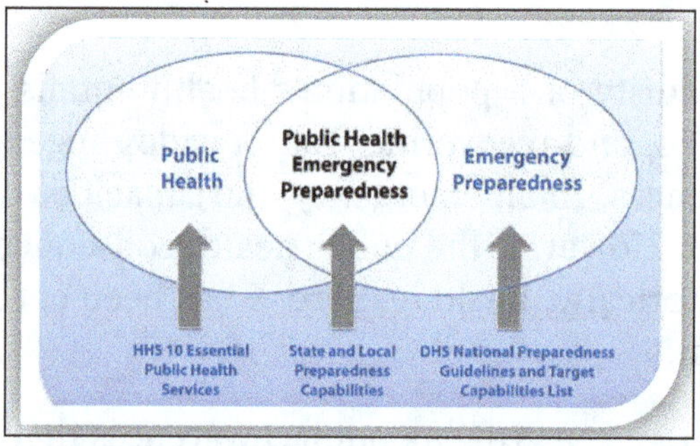

Public health emergency preparedness.

Public health emergency functions: Public health functions during emergencies must be derived and should support the accomplishment of the 10 essential public health services. Those functions are different for each phase of the disaster cycle. These which are related to the disaster prevention and preparedness are shown in the table below.

Public health emergency capabilities: State and local public health departments are first responders to public health incidents so, their preparedness capacity should be strengthened. Public health preparedness capabilities would assist public health departments in developing annual and long-term preparedness plans and would guide their preparedness strategies and investments.

The public health preparedness capabilities should represent a national public health standard for state and local preparedness of the health departments for responding to public health emergencies, in accordance with required public health functions during disasters. Each of the public health preparedness capabilities identifies priority resource elements that are relevant to both routine public health activities and essential public health services. While demonstrations of capabilities can be achieved through different means (e.g., exercises, planned events, and real incidents), jurisdictions are encouraged to use routine public health activities to demonstrate and evaluate their public health preparedness capabilities.

Table. Public health functions in the prevention/mitigation and preparedness phases during public health emergencies.

Emergency Prevention and Mitigation	Emergency Preparedness
• Conduct the risk assessment of health facilities. • Contribute in the community-disaster- risk-assessment process. • Monitor the risks and vulnerability of health facilities and the population over time. • Manage structural and nonstructural mitigations in health facilities. • Support public awareness on disaster risks and mitigation measures. • Ensure that the disaster risk reduction is considered in environmental policies and is operationalized by the relevant sectors. • Ensure that land-use regulations are in place to prevent population settlement and construction of health facilities in high-risk zones. • Ensure that all health facilities, other infrastructures, and houses are insured.	• Contribute in the community-disaster- risk-assessment process. • Conduct the risk assessment of health facilities. • Establish early-warning systems. • Develop the emergency response operations plan and conduct the associated activities. • Provide education and training of health authorities and personnel. • Conduct drills and exercises. • Monitor community preparedness for disasters. • Provide public awareness programs.

The US Department of Health and Human Services (HHS) identified the following 15 public health and health-care preparedness capabilities, divided in six core groups, as the basis for state and local public health and health-care preparedness. They are shown in table.

Table: Public health and health care preparedness capabilities:

Biosurveillance: • Public health surveillance and epidemiological investigation. • Public health laboratory testing.	Information management: • Emergency public information and warning information sharing.
Community resilience: • Community preparedness. • Community recovery.	Incident management: • Emergency operations coordination.
Countermeasures and mitigation: • Medical countermeasure dispensing. • Medical materiel management and distribution. • Non-pharmaceutical interventions. • Responder safety and health.	Surge management: • Fatality management. • Mass care. • Medical surge. • Volunteer management.

Each capability includes a definition of such capability and a list of the associated functions, performance measures, tasks, and resource considerations. The overall approach to implementation of the public health preparedness capabilities is to develop appropriate planning model i.e. to develop a preparedness plan.

Public health emergency preparedness plan: The public health emergency preparedness plan has the same structure and components like the disaster preparedness plan: risk assessment, planning, exercise, training and statutory enforcement. However, the public health emergency preparedness plan has some specificity as well. They are related to the needs for early warning and surveillance as key elements for assessing of the state of emergency.

The risk assessment is focused on the health system and the functioning of the health care facilities. Public health agencies are required to contribute actively to the process of community risk assessment by providing health-related indicators and information.

The next activity in the overall public health preparedness is development of an emergency response plan (ERP) and associated early warning and surveillance functions, training and exercises. An all-hazard/whole-health approach is recommended when planning for public health emergency. A whole-health approach entails the following components:

- Establishment of a unified preparedness and response platform for all categories of health risks.
- Inclusion of all health-related capabilities available by governmental and non-governmental organizations, the private sector, and the military.
- Assurance that all aspects of public health, including emergency medical care, pharmaceuticals, environmental health, communicable disease control, management of chronic diseases, mental health, reproductive health, and nutrition, are well addressed.

Activities related to the ERP include the development of command, control, and coordination mechanisms; surge capacity protocols; information management; and plans for communications, evacuation, public information, safety of medical staff, volunteers, and population, as well as the stockpiling of equipment and supplies. Formal institutional, legal, and budgetary capacities are needed to support the successful preparation and checking of the applicability of an ERP which would be performed during the exercises as a common component of the preparedness cycle.

Public health personnel must be integrated and participate with other response agencies during drills and exercises to better familiarize each stakeholder with their respective roles and abilities. In addition, this is the phase in which public health agencies would develop interagency agreements, memoranda of understanding (MOUs), and external support contracts.

The establishment of early-warning systems is a key component of the preparedness phase. Early warning is used to predict the likelihood of occurrence of public health emergency and offers the benefit of a rapid response and reduction in morbidity and mortality. In order to ensure a rapid and effective response by the EU to a wide range of emergencies, the Commission has put in place a number of early warning and rapid alert systems and open information sources (access to MedISys and other similar information sources) transmitting alerts and early warnings (EWRS, RAS BICHAT, RAS CHEM, RASFF, RAPEX and others relevant to the health crisis such as IHR).These systems are based on an information exchange network for receiving and triggering alerts, as well as exchanging other relevant information. Each of these systems covers a specific health threat field:

- The Early Warning and Response System (EWRS) for communicable diseases in the European Union was created by the European Commission to "ensure a rapid and effective response by the EU to events (including emergencies) related to communicable diseases.
- RAS BICHAT is the Rapid Alert System used for exchanging information on health threats due to deliberate release of chemical, biological, and radio-nuclear agents (notification of confirmed or suspected events, exchange of information and coordination of measures among partners).
- RAS CHEM is a rapid alert system used for the exchange of information on incidents including chemical agents relevant to terrorism and other events leading to release of chemicals, and consultation and coordination of counter-measures. This system should link the poison centers in the European Union, national chemical agencies and the Ministries of Health.

The EWRS for communicable diseases is the most developed system in case of preparedness to respond to the major public health threats even though the same efforts are given for chemical and radio- nuclear threats.

In response to the threat of communicable diseases, EU policy has focused on:

- Surveillance.
- Rapid detection.
- Rapid response.

Surveillance: Surveillance systems provide information for monitoring the trend of communicable diseases, early detection of outbreaks, and help to identify risk factors, and need for interventions. They provide information for priority setting, planning, implementation and resource allocation for preventive programs and for evaluation and control measures.

European Center for Communicable Diseases (ECDC) has developed indicator-based surveillance, the systematic ongoing collection, analysis, interpretation and dissemination of highly structured information ("indicators") for public health action. Indicator-based surveillance is complemented by event-based surveillance, detection, verification, analysis, assessment and further investigation of potential public health threats ("events"). Surveillance data collected at the European level are predominantly case-based and comprise demographic, clinical, epidemiological and laboratory information. ECDC collects analyses and disseminates surveillance data on 52 communicable diseases and related special health issues from all 28 European Union (EU) Member States and two of the three remaining European Economic Area (EEA) countries (Iceland and Norway).

Even though surveillance is very important and powerful mechanism for following an emergency situation, mathematical models have been more and more frequently used in recent years. Such models can help in analyzing the spread and control of infectious diseases and other harmful agents as health threats and advise decision makers in the area of public health preparedness, risk assessment and crisis management.

The combined use of remote sensing technology, Geographic Information Systems (GIS), spatial statistical techniques and mathematical models can help us in:

- Modeling of the dispersion of the harmful agent in the environment.
- Modeling of the exposure of the population to the harmful agent.

Modeling the dispersion of chemical agents is more highly developed than modeling for infectious agents. The modeling of risk zones is currently one of the requirements in the Safety Plans for chemical facilities under the Seveso Directive. Modeling of radiation dispersion has been developed in the radiation protection field and several activities have been undertaken at the different departments of the EC.

Rapid detection: It is essential with every public health threat to rapidly identify and confirm the agent involved. Every plan should address the identification of unknown agents, confirmation of known agents, and provision of surge capacity for a Member State facing a laboratory burden. In the laboratory domain, the plans should meet the above needs for both clinical and environmental sampling, with a coordination mechanism linking the actors if more than one is involved. In the following, these dual activities (analyzing clinical and environmental samples) should always be borne in mind.

For laboratory work, a structure should include procedures for laboratory reporting, confirmation of results (second lab, second country) and quality assurance. On the clinical side, clinicians should be able to identify the syndrome and a system should supply them quickly with the adequate guidelines.

For unknown agents, an international system has proved essential for rapid agreement on laboratory procedures and collating clinical data. For known agents of high threat potential, a secondary confirmation at an international level would improve trust in the diagnosis made. If there is a massive surge of samples in a single country, it will be essential to offer support by networking national and international laboratories.

These networks also must, if necessity, have common quality assurance schemes. Laboratory support would be available at Member State level, and for issues beyond national capacity or when national capacity is not available, cooperation between labs within the Community can be organized to optimize the use of pooled EU resources.

Hospital Preparedness Plan

An integral part of the overall public health emergency preparedness plan is the hospital preparedness plan. National/community plans would identify and verify the commonly agreed EU minimum requirements for health care facilities, taking into account such factors as the appropriateness and adequacy of physical facilities, organizational structures, human resources, and communication systems.

Emergency departments and intensive health care: Emergency departments in most hospitals might be confronted to a spontaneous influx of victims in case of a large scale incident (i.e. patients arriving at the emergency department without prior notice) necessitating triage arrangements outside the hospital. Generic preparedness plan ought to cover all aspects related to the work of Intensive Care Units (ICUs) in many different emergency scenarios. These have limited resources with expensive equipment that requires a significant amount of highly trained and capable staff for each patient and demand will often be much higher than the resources available. Every country needs plans for extending these resources to the maximum and possibly for stocks of additional equipment and pharmaceuticals to be used in an emergency.

Fatality management: Multi-sectoral (health, law enforcement, civil protection, forensic agencies) and multi-cultural approaches are mandatory in case of the fatality management during mass incidents. Plans will ensure that extra resources can be put in place when the number of cases overwhelms the system. The agents causing fatalities may demand special handling of bodies to protect personnel and ensure that the agent is not spread further in the environment.

Infection control/personal protective equipment: With many of the agents to consider when planning for a public health event, there are additional concerns about personal protective measures and infection control. Special precautions may be needed to reduce the likelihood of transmission and for certain diseases additional isolation requirements may be needed. In some instances it means using specialized equipment or pharmaceuticals for pre- and post-exposure prophylaxis. This must be appropriately regulated by the law in case of possible consequences of recommended/mandatory pre- and post-exposure measures.

Isolation/Quarantine procedures: For a number of agents, which are considered as public health hazards, isolation of patients is an important countermeasure as well as quarantine measures. It limits the spread of disease to personnel and other patients. For some diseases, isolation procedures have already been agreed, defining cases to be isolated and when they can be released; they need to be developed for others. In some areas specialized resources will be needed and national preparedness plan should include the amount required additional resources.

Decontamination: In some cases of chemical or radiological health emergency, the decontamination of the patients/victims should be necessary. Decontamination of exposed individuals prior to receiving them in the health care facility may be necessary to ensure the safety of patients and staff while providing care. Procedures to determine the need for decontamination and capabilities to carry out the requisite measures would be developed in the frame of the public health emergency preparedness plan.

Transport of samples and patients: Preparedness plans should pay special attention to transport of dangerous pathogens in the country and, if needed, out of the country as well.. Sometimes such transports are approved by competent authorities based on agreements between the laboratories. It is usual to inform the Customs on the type of transportation chosen to by the national laboratories in order to be prepared for public health emergencies. The relevant legislation should be put in place for covering the issues related to the transport of patient and environmental samples in public health emergencies. Local and national legislation on the transport of dangerous substances should be regularly adapted to the evolving scientific and technological circumstances, as reported by international organizations.

Ethical implications of countermeasures: Ethical issues are closely related to the legal issues as mentioned above and are part of the framework needed to assess the cultural acceptability of measures such as quarantine and selective immunization of pre-defined risk groups. Ethical aspects arise when there is a shortage of means for the whole population and only selected groups can get access to available resources. National preparedness plans have to include a guideline for the ethical framework of responses to public health crises. National plans *have to* include a verification process to ensure that the ethical aspects of policy decisions during an outbreak provide a balance between individual rights and population rights.

Infrastructure safety. The health care infrastructure has become increasingly dependent on other infrastructure systems, so preparedness within the health care sector must cover those dependencies. Especially systems for supply of electric power, water, medical gas and information systems should be taken into account. Also, special attention should be paid to the protection of hospitals from contamination by toxic chemicals and the spreading of contagious substances, by adjusting ventilation systems etc. Other aspects of safe hospitals are preparedness for fire and flooding.

The risk and vulnerability analysis should include preparedness for disruption in various infrastructure systems, externally and internally (safe hospital aspects) and need to provide services in a timely and 24-hour manner. Examples include the Kaiser Permanente Hazard Vulnerability Assessment (HVA) tool, which allows staff in medical centers to prioritize potential events based on calculated risk, by using probability and severity. The Hospital Safety Index (HSI), developed by the WHO assesses the safety level of a hospital in three main components (functional capacity, structural safety, and nonstructural safety).

The Hazard Vulnerability Analysis (HVA) provides a systematic approach to recognizing hazards that may affect demand for the hospital's services or its ability to provide those services. The risks associated with each hazard (natural, man-made, and hazardous substances) are analyzed to prioritize planning, mitigation, response, and recovery activities. Kaiser Permanente has developed a Hazard Vulnerability Analysis tool which is available for download as a planning resource.

The Hospital Safety Index occupies a central place in the local, national, and global efforts to improve the functioning of hospitals in emergencies and disasters. A safe hospital is a facility which services remain accessible and functioning at a maximum capacity, and with the same infrastructure, before, during and immediately after the impact of emergencies and disasters. A key element of progress towards safe hospitals has been the development and application of the Hospital Safety Index – a rapid and low-cost diagnostic tool for assessing the probability that a hospital will remain operational in emergencies and disaster. The original Hospital Safety Index was developed by PAHO and WHO and was released in 2008. Subsequently, the Hospital Safety Index tool has been used to assess the safety of more than 3,500 facilities and has been adopted and implemented by many countries.

Levels of Preparedeness Activities

Preparedness carries an unfair stigma associated with doomsday preppers and fallout shelters. The cold hard truth is that even the most basic preparedness plan can ultimately save not only the lives of your family members, but the lives of other people. You don't need a bug out vehicle or a bunker filled with non-perishables to be ready for a crisis. Often, preparedness just comes down to your ability to live comfortably if your power goes out for more than a day. We have consulted preparedness experts and have broken down preparedness into four levels in ascending order.

Level One: Basic Preparedness

Preparedness level: 3-5 days of resources:

This is a very simple way to make sure you are prepared for evacuations, multiple day long power outages or any sudden inconvenience that will alter your everyday comforts of home. Your life may not be in danger, but you certainly need this level of preparedness to avoid major headaches and maintain a relative level of comfort in extraordinary circumstances. Every home should include a surplus of these items.

Resources Needed

- Heat source: A fireplace or non-electric, safe and contained source of heat to keep you warm in your home.
- A first aid kit: Not just bandaids and antiseptic cream. Include multiple types of bandages (gauze, butterfly, etc.), sunscreen, pain relievers, hand sanitizer, epi-pens, etc.
- Drinking water: The rule of thumb is a gallon per person, per day, for 3-5 days.
- Communication: Two way radios to reach neighbors and nearby family members.
- Batteries.
- A full or near-full tank of gas in your car if an evacuation is imminent.
- Canned goods and non-perishable food. Enough to last everyone in the home 3-5 days.
- Prescription medicine: Keep an extra week's supply on hand if possible.
- Battery powered emergency NOAA AM/FM radio.
- A flashlight.
- A small supply of cash.
- Candles.
- An emergency savings fund reserved for 3-5 days at hotels, food and traveling.

Level Two: Extended Preparedness

Preparedness level: 1-3 weeks of resources:

You and your family will likely not be able to receive immediate assistance from emergency services and law enforcement for several days, perhaps weeks. This level of emergency, like an ice storm or hurricane, may have caused roadways to be cut-off and disrupted your community's supply chain of food, oil, water and other vital resources.

What you Need

- A go-bag: Are you ready to leave your home at a moment's notice if it becomes uninhabitable? Pack food (granola, nuts, energy bars), water, medicine and small tools to last you 24-48 hours outside of your home while you seek safety.
- A generator: Keeping the power on in your home to keep it lit and fulfill small energy needs like lighting and small appliances.
- Self defense: Unfortunately, looting and burglary are big problems in the aftermath of emergencies. Keep your family and your property protected.
- Cooking methods: Dutch ovens and cast iron over fire or canned heat can give you reprieve from eating every meal from a can.
- Water purification systems: Tablets, filtered straws or boil methods.
- Sanitation items: One month supply of toilet paper, feminine hygiene products, etc.
- An emergency fund of one month's living expenses, as well as a decent amount of cash.

Level Three: Advanced Preparedness

Preparedness level: At least one month of resources:

Imagine New Orleans after Hurricane Katrina. Food, oil and water supply chains were cut off and the cleanup effort took months. This left many residents to fend for themselves which led to looting, rampant crime and lawlessness. Law enforcement and emergency services will be spread thin and response times will be severely delayed. Martial law will be in effect and lawlessness would be a genuine threat. Would you be ready for this?

What you Need

- High powered communication: Can you reach outside of the affected area (Over 10 miles)? While walkie talkies advertise ranges of 20+ miles, they are most effective in ranges that are less than two miles. Repeater enabled GMRS radios are easy to operate and can carry messages over a long distance to anyone who might be able to help.
- Rechargeable batteries and other renewable energy sources.
- Barter items: Make sure you have a surplus of items available to trade or even share with neighbors, especially food and batteries.
- Waste disposal methods.

- An emergency fund of at least 3-6 months of living expenses. When will you be able to work again? Will you still have a job?

Level Four: Catastrophic Preparedness

Preparedness level: Potentially permanent:

In the extremely unlikely event of total economic and societal collapse happening, you will want to be at this level. This level is extremely complex and advanced.

Resources Needed

- A seed vault and livestock. You will be growing and butchering your own food at this point.
- Renewable water source, such as a well or rainwater barrel.
- Water purification method.
- Renewable energy source(s).
- Large surplus of self defense items.
- Heavy duty first aid kit, ability to perform minor surgery.
- Herbal medicine knowledge.
- Foraging knowledge.
- A food production system.

As you can see, the levels of preparedness vary. Even at the most basic level of preparedness, you will be able to keep your family safe. However, you can also take matters into your own hands and build a rock-solid preparedness plan and prepare for catastrophic disasters and circumstances that are likely to never happen.

Steps in Disaster Preparedness and Response Planning

Risk Assessment

- The initial step of disaster preparedness and response planning is to assess the risk.
- All the assessment regarding the hazard and vulnerabilities will be done in this stage.
- All the baseline information will be collected.
- The past incidence and results of the disaster are reviewed.

Developing the Workforce

- Once the assessment is done, workforce is prepared.
- The required workforce are recruited.
- Capacity building of the workforce is done through training and workshops.
- Roles and responsibilities of the people and institutions are made clear.

Make Critical Position Available

- All the available places are identified.
- The local people need to be made aware about the safe places.
- The places where camps will be conducted, needs to be made known to general public.

Update and Alert

- Updates and alertness are necessary.
- Alertness can be provided prior to the unset of the disaster and emergency.
- Timely alert helps in evacuation, reducing the loss of life.
- Updating the preparedness plan needs to be done time and again.
- Early warning system should also be updated and upgraded.

Test the Plan

- It is very necessary to test the plan.
- It helps in identifying loop holes in the plans.
- Drills can be conducted in the community by the help of the community and other enterprises.

Preparedness as a whole Community Activity

In the United States, prior to marri, preparedness was largely viewed as the responsibility of first responders and other emergency services. In the aftermath of Katrina, it became evident that first responders can and will become overwhelmed in a large-scale disaster; unable to effectively respond to the emergency. Individual preparedness must be undertaken; ready to handle any disaster. The idea of whole community preparedness, is "By working together, everyone can keep the nation safe from harm and resilient when struck by hazards, such as natural disasters, acts of terrorism, and pandemics."

The United Emergency Management Agency (FEMA), individuals, families, businesses, faith-based and community groups profitable groups, schools and academia, media outlets, and all level of governments must take an active role in preparedness efforts. A disaster will affect the whole community, so everyone must be ready, by making a plan, being informed, and taking action to mitigate the effects of future disasters.

Disaster Response

A mobile emergency operations center.

Disaster response is the second phase of the disaster management cycle. It consists of a number of elements, for example; warning/evacuation, search and rescue, providing immediate assistance, assessing damage, continuing assistance and the immediate restoration or construction of infrastructure (i.e. provisional storm drains or diversion dams). The aim of emergency response is to provide immediate assistance to maintain life, improve health and support the morale of the affected population. Such assistance may range from providing specific but limited aid, such as assisting refugees with transport, temporary shelter, and food, to establishing semi-permanent settlement in camps and other locations. It also may involve initial repairs to damaged or diversion to infrastructure.

The focus in the response phase is on putting people safe, prevent need disasters and meeting the basic needs of the people until more permanent and sustainable solutions can be found. The main responsibility to address these needs and respond to a disaster lies with the government or governments in whose territory the disaster has occurred. In addition, Humanitarian organizations are often strongly present in this phase of the disaster management cycle, particularly in countries where the government lacks the resources to respond adequately to the needs.

Disaster Response Planning

The United States National Fire Protection Association (NFPA) 1600 Standard specify elements of an emergency response, as: defined responsibilities; specific actions to be taken (which must include protective actions for life safety); and communication directives. Within the standard, NFPA recognize that disasters and day-to-day emergencies are characteristically different. Nevertheless, the prescribed response elements are the same.

In support of the NFPA standard, Statoil's practical application of emergency response is across three distinct "lines" that incorporate NFPA's elements. Line 1 is responsible for the operational management of an incident; line 2, typically housed off-site, is responsible for tactical guidance and additional resource management. Finally, in the case of major incidents, line 3 provides strategic guidance, group resource management, and government and media relations.

While it is impossible to plan for every disaster, crisis or emergency, the Statoil investigation into the terrorist attacks on in Amenas place emphasis on the importance of having a disaster response. The report concludes that a disaster response framework may be utilized in an array of disaster situations, such as that at in Amenas.

Disaster Response Technologies

Smart Emergency Response System (SERS) prototype was built in the SmartAmerica Challenge 2013-2014, a United States government initiative. SERS has been created by a team of nine organizations led by MathWorks. The project was featured at the White House in June 2014 and described by Todd Park (U.S. Chief Technology Officer) as an exemplary achievement.

The SmartAmerica initiative challenges the participants to build cyber-physical systems as a glimpse of the future to save lives, create jobs, foster businesses, and improve the economy. SERS primarily saves lives. The system provides the survivors and the emergency personnel with information to locate and assist each other during a disaster. SERS allows to submit help requests to a MATLAB-based mission center connecting first responders, apps, search-and-rescue dogs, a 6-feet-tall humanoid, robots, drones, and autonomous aircraft and ground vehicles. The command and control center optimizes the available resources to serve every incoming requests and generates an action plan for the mission. The Wi-Fi network is created on the fly by the drones equipped with antennas. In addition, the autonomous rotorcrafts, planes, and ground vehicles are simulated with Simulink and visualized in a 3D environment (Google Earth) to unlock the ability to observe the operations on a mass scale.

The International Charter Space and Major Disasters provides for the charitable retasking of satellite assets, providing coverage from 15 space agencies, etc. which is wide albeit contingent. It focuses on the beginning of the disaster cycle, when timely data is of the essence.

Digital technologies are increasingly being used in humanitarian action, they have shown to improve the health and recovery of populations affected by both natural and man-made disasters. They are used in humanitarian response to facilitate and coordinate aid in various stages including preparedness, response, and recovery from emergencies. More specifically, mobile health (mHealth), which is defined as the use of communication devices such as mobile phones for the purpose of health services information. Nowadays, millions of people use mobile phones as a means of daily communication and data transference, out of which 64% live in developing countries. One of the most important characteristics of disasters are the harms caused to infrastructures, accessibility issues, and an exponential need of medical and emergency services. In such situations, the use of mobile phones for mHealth can be vital, especially when other communication infrastructures are hindered. In such conditions, the abundance of mobile technology in developing countries provide the opportunity to be harnessed for helping victims and vulnerable people.

Mobile health information technology platforms, in the acute phase of disaster response, create a common operational framework that improves disaster response by standardizing data acquisition, organizing information storage, and facilitating communication among medical staff. One of the challenges in disaster response is the need of pertinent, effective and continuous analysis of the situation and information in order to evaluate needs and resources. mHealth has been shown to provide effective disaster preparedness with real time collection of medical data as well as helping identify and create needs assessments during disasters. Using mobile technology in heath has set the stage for the dynamic organization of medical resources and promotion of patient care done through quick triage, patient tracking, and documentation storage and maintenance.

Managing an effective and influential response requires cooperation, which is also facilitated through mHealth. A retrospective study demonstrated that applying mHealth can lead to up to 15% decrease of unnecessary hospital transfers during disasters. In addition, they provide field hospital administrators with real-time census information essential for planning, resource allocation, inter-facility patient transfers, and inter-agency collaboration. mHealth technology systems can improve post-operative care and patient handoffs between volunteer providers. Data entry with mobile devices is now widely used to facilitate the registration of displaced individuals, to conduct surveys, identify those in need of assistance, and to capture data on issues such as food security, vaccination rates, and mortality.

Above all, mHealth can harness the power of information to improve patient outcomes. Efforts lead by the Harvard Humanitarian Initiative and Operational Medicine Institute during the Haiti earthquake resulted in the creation of a web-based mHealth system that created a patient log of 617 unique entries used by on-the-ground medical providers and field hospital administrators. This helped facilitate provider triage, improve provider handoffs, and track vulnerable populations such as unaccompanied minors, pregnant women, traumatic orthopedic injuries and specified infectious diseases. Also, during the Haiti earthquake, the International Red Crescent sent more than 45 million SMSs to Viole mobile phone users. This resulted in 95% of the receiver reporting they had gained useful information, and out of these 90% reported the SMS helped in their preparedness.

Objectives of Response

As indicated by the situation, response activities include:

- Applying intelligence and other information to lessen the effects or consequences of an incident.
- Increasing security operations.
- Continuing investigations into the nature and source of the threat.
- Ongoing public health and agricultural surveillance and testing processes, immunizations, isolation, or quarantine.
- Specific law enforcement operations aimed at preempting, interdicting, or disrupting illegal activity, and apprehending actual perpetrators and bringing them to justice.
- Restoring critical infrastructure (e.g., utilities).
- Ensuring continuity of critical services (e.g., law enforcement, public works). In other words, response involves putting preparedness plans into action.

Emergency Response Plan

The actions taken in the initial minutes of an emergency are critical. A prompt warning to employees to evacuate, shelter or lockdown can save lives. A call for help to public emergency services that provides full and accurate information will help the dispatcher send the right responders and equipment. An employee trained to administer first aid or perform CPR can be lifesaving. Action by employees with knowledge of building and process systems can help control a leak and minimize damage to the facility and the environment.

The first step when developing an emergency response plan is to conduct a risk assessment to identify potential emergency scenarios. An understanding of what can happen will enable you to determine resource requirements and to develop plans and procedures to prepare your business. The emergency plan should be consistent with your performance objectives.

At the very least, every facility should develop and implement an emergency plan for protecting employees, visitors, contractors and anyone else in the facility. This part of the emergency plan is called "protective actions for life safety" and includes building evacuation ("fire drills"), sheltering from severe weather such as tornadoes, "shelter-in-place" from an exterior airborne hazard such as a chemical release and lockdown. Lockdown is protective action when faced with an act of violence.

When an emergency occurs, the first priority is always life safety. The second priority is the stabilization of the incident. There are many actions that can be taken to stabilize an incident and minimize potential damage. First aid and CPR by trained employees can save lives. Use of fire extinguishers by trained employees can extinguish a small fire. Containment of a small chemical spill and supervision of building utilities and systems can minimize damage to a building and help prevent environmental damage.

Some severe weather events can be forecast hours before they arrive, providing valuable time to protect a facility. A plan should be established and resources should be on hand, or quickly, available to prepare a facility. The plan should also include a process for damage assessment, salvage, protection of undamaged property and cleanup following an incident. These actions to minimize further damage and business disruption are examples of property conservation.

Protective Actions for Life Safety

When there is a hazard within a building such as a fire or chemical spill, occupants within the building should be evacuated or relocated to safety. Other incidents such as a bomb threat or receipt of a suspicious package may also require evacuation. If a tornado warning is broadcast, everyone should be moved to the strongest part of the building and away from exterior glass. If a transportation accident on a nearby highway results in the release of a chemical cloud, the fire department may warn to "shelter-in-place." To protect employees from an act of violence, "lockdown" should be broadcast and everyone should hide or barricade themselves from the perpetrator.

Protective actions for life safety include:

- Evacuation.
- Sheltering.
- Shelter-in-Place.
- Lockdown.

Your emergency plan should include these protective actions. If you are a tenant in multi-tenanted building, coordinate planning with the building manager.

Evacuation

Prompt evacuation of employees requires a warning system that can be heard throughout the building. Test your fire alarm system to determine if it can be heard by all employees. If there is no fire alarm system, use a public address system, air horns or other means to warn everyone to evacuate. Sound the evacuation signal during planned drills so employees are familiar with the sound.

Make sure that there are sufficient exits available at all times:

- Check to see that there are at least two exits from hazardous areas on every floor of every building. Building or fire codes may require more exits for larger buildings.

- Walk around the building and verify that exits are marked with exit signs and there is sufficient lighting so people can safely travel to an exit. If you find anything that blocks an exit, have it removed.
- Enter every stairwell, walk down the stairs, and open the exit door to the outside. Continue walking until you reach a safe place away from the building. Consider using this safe area as an assembly area for evacuees.

Appoint an evacuation team leader and assign employees to direct evacuation of the building. Assign at least one person to each floor to act as a "floor warden" to direct employees to the nearest safe exit. Assign a backup in case the floor warden is not available or if the size of the floor is very large. Ask employees if they would need any special assistance evacuating or moving to shelter. Assign a "buddy" or aide to assist persons with disabilities during an emergency. Contact the fire department to develop a plan to evacuate persons with disabilities.

Have a list of employees and maintain a visitor log at the front desk, reception area or main office area. Assign someone to take the lists to the assembly area when the building is evacuated. Use the lists to account for everyone and inform the fire department whether everyone has been accounted for. When employees are evacuated from a building, OSHA regulations require an accounting to ensure that everyone has gotten out safely. A fire, chemical spill or other hazard may block an exit, so make sure the evacuation team can direct employees to an alternate safe exit.

Sheltering

If a tornado warning is broadcast, a distinct warning signal should be sounded and everyone should move to shelter in the strongest part of the building. Shelters may include basements or interior rooms with reinforced masonry construction. Evaluate potential shelters and conduct a drill to see whether shelter space can hold all employees. Since there may be little time to shelter when a tornado is approaching, early warning is important. If there is a severe thunderstorm, monitor news sources in case a tornado warning is broadcast. Consider purchasing an Emergency Alert System radio - available at many electronic stores. Tune in to weather warnings broadcast by local radio and television stations. Subscribe to free text and email warnings, which are available from multiple news and weather resources on the Internet.

Shelter-in-Place

A tanker truck crashes on a nearby highway releasing a chemical cloud. A large column of black smoke billows into the air from a fire in a nearby manufacturing plant. If, as part of this event, an explosion, or act of terrorism has occurred, public emergency officials may order people in the vicinity to "shelter-in-place." You should develop a shelter-in-place plan. The plan should include a means to warn everyone to move away from windows and move to the core of the building. Warn anyone working outside to enter the building immediately. Move everyone to the second and higher floors in a multistory building. Avoid occupying the basement. Close exterior doors and windows and shut down the building's air handling system. Have everyone remain sheltered until public officials broadcast that it is safe to evacuate the building.

Lockdown

An act of violence in the workplace could occur without warning. If loud "pops" are heard and gunfire is suspected, every employee should know to hide and remain silent. They should seek refuge in a room, close and lock the door, and barricade the door if it can be done quickly. They should be trained to hide under a desk, in the corner of a room and away from the door or windows. Multiple people should be trained to broadcast a lockdown warning from a safe location.

Incident Stabilization

Stabilizing an emergency may involve many different actions including: firefighting, administering medical treatment, rescue, containing a spill of hazardous chemicals or handling a threat or act of violence. When you dial emergency helpline number you expect professionals to respond to your facility. Depending upon the response time and capabilities of public emergency services and the hazards and resources within your facility, you may choose to do more to prepare for these incidents. Regulations may require you to take action before emergency services arrive.

If you choose to do nothing more than call for help and evacuate, you should still prepare an emergency plan that includes prompt notification of emergency services, protective actions for life safety and accounting of all employees.

Developing the Emergency Plan

Developing an emergency plan begins with an understanding of what can happen. Review your risk assessment. Consider the performance objectives that you established for your program and decide how much you want to invest in planning beyond what is required by regulations.

Assess what resources are available for incident stabilization. Consider internal resources and external resources including public emergency services and contractors. Public emergency services include fire departments that may also provide rescue, hazardous materials and emergency medical services. If not provided by your local fire department, these services may be provided by another department, agency or even a private contractor. Reach out to local law enforcement to coordinate planning for security related threats.

Document available resources. Determine whether external resources have the information they would need to handle an emergency. If not, determine what information is required and be sure to document that information in your plan.

Warning, Notifications and Communications

Plans should define the most appropriate protective action for each hazard to ensure the safety of employees and others within the building. Determine how you will warn building occupants to take protective action. Develop protocols and procedures to alert first responders including public emergency services, trained employees and management. Identify how you will communicate with management and employees during and following an emergency.

Roles and Responsibilities for Building Owners and Facility Managers

Assign personnel the responsibility of controlling access to the emergency scene and for keeping people away from unsafe areas. Others should be familiar with the locations and functions of controls for building utility, life safety and protection systems. These systems include ventilation, electrical, water and sanitary systems; emergency power supplies; detection, alarm, communication and warning systems; fire suppression systems; pollution control and containment systems; and security and surveillance systems. Personnel should be assigned to operate or supervise these systems as directed by public emergency services if they are on-site.

Site and Facility Plans and Information

Public emergency services have limited knowledge about your facility and its hazards. Therefore, it is important to document information about your facility. That information is vital to ensure emergency responders can safely stabilize an incident that may occur. Documentation of building systems may also prove valuable when a utility system fails—such as when a water pipe breaks and no one knows how to shut off the water.

Compile a site-plan and plans for each floor of each building. Plans should show the layout of access roads, parking areas, buildings on the property, building entrances, the locations of emergency equipment and the locations of controls for building utility and protection systems. Instructions for operating all systems and equipment should be accessible to emergency responders.

Provide a copy of the plan to the public emergency services that would respond to your facility and others with responsibility for building management and security. Store the plan with other emergency planning information such as chemical Material Safety Data Sheets (MSDS), which are required by Hazard Communication or "right to know" regulations.

Training and Exercises

Train personnel so they are familiar with detection, alarm, communications, warning and protection systems. Review plans with staff to ensure they are familiar with their role and can carry out assigned responsibilities. Conduct evacuation, sheltering, sheltering-in-place and lockdown drills so employees will recognize the sound used to warn them and they will know what to do. Facilitate exercises to practice the plan, familiarize personnel with the plan and identify any gaps or deficiencies in the plan.

10 Steps for Developing the Emergency Response Plan

- Review performance objectives for the program.
- Review hazard or threat scenarios identified during the risk assessment.
- Assess the availability and capabilities of resources for incident stabilization including people, systems and equipment available within your business and from external sources.
- Talk with public emergency services (e.g., fire, police and emergency medical services) to determine their response time to your facility, knowledge of your facility and its hazards and their capabilities to stabilize an emergency at your facility.
- Determine if there are any regulations pertaining to emergency planning at your facility; address applicable regulations in the plan.
- Develop protective actions for life safety (evacuation, shelter, shelter-in-place, lockdown).
- Develop hazard and threat-specific emergency procedures using the Emergency Response Plan Template for Businesses.
- Coordinate emergency planning with public emergency services to stabilize incidents involving the hazards at your facility.
- Train personnel so they can fulfill their roles and responsibilities.
- Facilitate exercises to practice your plan.

Hazards to Consider When Developing the Emergency Plan

Natural Hazards

- Geological Hazards:
    - Earthquake.

- Tsunami.
- Volcano.
- Landslide, mudslide, subsidence.

- Meteorological Hazards:
  - Flood, flash flood, tidal surge.
  - Water control structure/dam/levee failure.
  - Drought.
  - Snow, ice, hail, sleet, arctic freeze.
  - Windstorm, tropical cyclone, hurricane, tornado, dust storm.
  - Extreme temperatures (heat, cold).
  - Lightning strikes (wildland fire following).
- Biological Hazards:
  - Foodborne illnesses.
  - Pandemic/Infectious/communicable disease (Avian flu, H1N1, etc.)

Human-caused Events

Accidental

- Hazardous material spill or release.
- Nuclear power plant incident (if located in proximity to a nuclear power plant).
- Explosion/Fire.
- Transportation accident.
- Building/structure collapse.
- Entrapment and or rescue (machinery, confined space, high angle, water).
- Transportation Incidents (motor vehicle, railroad, watercraft, aircraft, pipeline).

Intentional

- Robbery.
- Lost person, child abduction, kidnap, extortion, hostage incident, workplace violence.

- Demonstrations, civil disturbance.
- Bomb threat, suspicious package.
- Terrorism.

Technology caused Events

- Utility interruption or failure (telecommunications, electrical power, water, gas, steam, HVAC, pollution control system, sewerage system, other critical infrastructure).
- Cyber security (data corruption/theft, loss of electronic data interchange or ecommerce, loss of domain name server, spyware/malware, vulnerability exploitation/botnets/hacking, denial of service).

Property Conservation

Taking action before a forecast event, such as a severe storm, can prevent damage. Prompt damage assessment and cleanup activities following the storm can minimize further damage and business disruption. These actions are considered "property conservation"—an important part of the emergency response plan. Much of the following guidance is directed to building owners and facility managers. However, tenants should also develop a plan in coordination with building owners and managers as well as public authorities.

Preparing a Facility for a Forecast Event

Body copy: Actions to prepare a facility for a forecast event depend upon the potential impacts from the hazards associated with the event. Conduct a risk assessment to identify severe weather hazards including winter storms, arctic freeze, tropical storm, hurricane, flooding, storm surge, severe thunderstorm, tornado and high winds. Also consider non-traditional hazards, such as a planned event involving a large crowd.

Property conservation actions should focus on protection of the building and valuable machinery, equipment and materials inside. Potential damage may be prevented or mitigated by inspecting the following building features, systems and equipment:

- Windows and doors.
- Roof flashing, covering and drainage.
- Exterior signs.
- Mechanical equipment, antennas and satellite dishes on rooftops.
- Outside storage, tanks and equipment.

- Air intakes.
- High value machinery.
- Sensitive electronic equipment including information technology and process controllers.

The review of building components may also identify opportunities for longer-term mitigation strategies.

Property conservation activities for specific forecast events include the following:

- Winter storm: Keep building entrances and emergency exits clear; ensure there is adequate fuel for heating and emergency power supplies; monitor building heat, doors and windows to prevent localized freezing; monitor snow loading and clear roof drains.
- Tropical storms and hurricanes: Stockpile and pre-cut plywood to board up windows and doors (or install hurricane shutters); ensure there is sufficient labor, tools and fasteners available; inspect roof coverings and flashing; clear roof and storm drains; check sump and portable pumps; backup electronic data and vital records off-site; relocate valuable inventory to a protected location away from the path of the storm.
- Flooding: Identify the potential for flooding and plan to relocate goods, materials and equipment to a higher floor or higher ground. Clear storm drains and check sump and portable pumps. Raise stock and machinery off the floor. Prepare a plan to use sandbags to prevent water entry from doors and secure floor drains.

Salvage and Actions to Prevent Further Damage Following an Incident

Separating undamaged goods from water-soaked goods is an example of salvage. Covering holes in a roof or cleaning up water and ventilating a building are also part of property conservation. The property conservation plan should identify the resources needed to salvage undamaged good and materials; make temporary repairs to a building; clean up water, smoke and humidity; and prepare critical equipment for restart.

Resources for property conservation include the following:

- Water vacuums and tools to remove water.
- Fans to remove smoke and humidity.
- Tarpaulins or plywood to cover damaged roofs or broken windows.
- Plastic sheeting to cover sensitive equipment.

Compile an inventory of available equipment, tools and supplies and include it with the emergency response plan. Identify precautions for equipment exposed to water or high humidity and procedures for restarting machinery and equipment.

Identify contractors that may be called to assist with clean up and property conservation efforts. Keep in mind that competition for contractors, labor, materials and supplies prior to a forecast storm or following a regional disaster may be intense. Plan ahead and secure contractors and other resources in advance.

References

- Disaster-preparedness- 301553185: researchgate.net, Retrieved 05 August, 2020
- Four-levels-preparedness: midlandusa.com, Retrieved 05 January, 2020
- Disaster-preparedness-and-response-planning: publichealthnotes.com, Retrieved 25 June, 2020
- TheFivePhasesofEmergencyManagement, EmergencyManagement: stlouisco.com, Retrieved 28 April, 2020
- Emergency: ready.gov, Retrieved 16 May, 2020

4

Recovery and Mitigation

Recovery deals with the rebuilding and repair of destroyed infrastructure for restoring community's systems and activities to normal. Mitigation lessens the impact of disasters and reduces the loss of life and property. This chapter discusses the concepts associated with recovery and mitigation in emergency management in detail.

Recovery

Community recovery from disasters can be a complex and often lengthy process, with different communities recovering at different rates. The recovery element of the comprehensive approach to disaster management – prevention, preparedness, response and recovery (PPRR) – can be the most complicated and protracted.

The best outcomes are achieved by ensuring recovery strategies align with community need and are led by the affected community.

This requires a collaborative, coordinated, adaptable and scalable approach where the responsibility for disaster recovery is shared among all sectors of the community including individuals, families, community groups, businesses and all levels of government.

A community-led approach supports the rapid restoration of services essential to human wellbeing and presents an opportunity to build resilience and improve community circumstances and preparedness beyond their pre-disaster status. Queensland takes an all hazards approach to recovery, as identified in the Act.

Principles

The following principles underpin all recovery planning and operations in Queensland:

- Understanding the context:
    - Acknowledge existing strengths and capacity, including past experiences;

  - Appreciate the risks and stressors faced by the community;
  - Be respectful of and sensitive to the culture and diversity of the community;
  - Support those who may be facing vulnerability;
  - Recognise the importance of the environment to people and to their recovery;
  - Be acknowledged as requiring a long term, sustained effort as needed by the community;
  - Acknowledge the impact upon the community may extend beyond the geographical boundaries where the disaster occurred.
- Recognising complexity:
  - Disasters lead to a range of effects and impacts that require a variety of approaches; they can also leave long-term legacies;
  - Information on impacts is limited at first and changes over time;
  - Affected individuals and the community have diverse needs, wants and expectations, which can evolve rapidly;
  - Responsive and flexible action is crucial to address immediate needs;
  - Existing community knowledge and values may challenge the assumptions of those outside of the community;
  - Conflicting knowledge, values and priorities among individuals, the community and organisations may create tensions;
  - Emergencies create stressful environments where grief or blame may also affect those involved;
  - Over time, appropriate support for individuals and communities, from within and outside, can cultivate hope and individual and collective growth.
- Using local, community-led approaches:
  - Assist and enable individuals, families and the community to actively participate in their own recovery;
  - Recognise that individuals and the community may need different levels of support at various times;
  - Be guided by the communities priorities;
  - Channel effort through pre-identified and existing community assets, including local knowledge, existing community strengths and resilience;

  - Build collaborative partnerships between the community and those involved in the recovery process;
  - Recognise that new community leaders often emerge during and after a disaster, who may not hold formal positions of authority; and Recognise that different communities may choose different paths to recovery.
- Ensuring coordination of all activities:
  - Have clearly articulated and shared goals based on desired outcomes;
  - Be flexible, taking into account changes in community needs or stakeholder expectations.
  - Be guided by those with experience and expertise, using skilled, authentic and capable community leadership;
  - Be at the pace desired by the community, and seek to collaborate and reconcile different interests and time frames;
  - Reflect well-developed community planning and information gathering before, during and after a disaster;
  - Have clear decision-making and reporting structures and sound governance, which are transparent and accessible to the community;
  - Demonstrate an understanding of the roles, responsibilities and authority of organisations involved and coordinate across agencies to ensure minimal service provision disruption;
  - Be part of an emergency management approach that integrates with response operations and contributes to future prevention and preparedness; and
  - Be inclusive, availing of and building upon relationships created before, during and after the emergency.
- Employing effective communication:
  - Recognise that communication should be two way, and that input and feedback should be encouraged;
  - Ensure that information is accessible to audiences in diverse situations, addresses a variety of communication needs, and is provided through a range of communication channels and networks;
  - Establish mechanisms for coordinated and consistent communications between all service providers, organisations and individuals and the community;

  - Ensure that all communication is relevant, timely, clear, accurate, targeted, credible and consistent; and
  - Identify trusted sources of information and repeat key recovery messages to enable greater community confidence and receptivity.
- Acknowledging and building capacity:
  - Assess capability and capacity requirements before, during and after a disaster;
  - Support the development of self-reliance, preparation and disaster mitigation;
  - Quickly identify and mobilise community skills, strengths and resources;
  - Develop networks and partnerships to strengthen capacity, capability and resilience;
  - Provide opportunities to share, transfer and develop knowledge, skills and training;
  - Recognise that resources can be provided by a range of partners and from community networks;
  - Acknowledge that existing resources may be stretched, and that additional resources may be sought;
  - Understand that additional resources may only be available for a limited period, and that sustainability may need to be addressed;
  - Understand when and how to step back, while continuing to support individuals and the community as a whole to be more self-sufficient when they are ready;
  - Be evaluated to provide learning for future disaster and improved resilience.

Disaster Recovery involves a set of policies, tools and procedures to enable the recovery or continuation of vital technology infrastructure and systems following a natural or human-induced disaster. Disaster recovery focuses on the IT or technology systems supporting critical business functions, as opposed to business continuity, which involves keeping all essential aspects of a business functioning despite significant disruptive events. Disaster recovery can therefore be considered as a subset of business continuity.

IT Service Continuity

IT Service Continuity (ITSC) is a subset of Business Continuity Planning (BCP) and encompasses IT disaster recovery planning and wider IT resilience planning. It also incorporates those elements of IT infrastructure and services which relate to communications such as (voice) telephony and data communications.

The ITSC Plan reflects Recovery Point Objective (RPO - recent transactions) and Recovery Time Objective (RTO - time intervals).

Principles of Backup Sites

Planning includes arranging for backup sites, be they hot, warm, cold or standby sites with hardware as needed for continuity.

In 2008 the British Standards Institution launched a specific standard connected and supporting the Business Continuity Standard BS 25999 titled BS25777 specifically to align computer continuity with business continuity. This was withdrawn following the publication in March 2011 of ISO/IEC 27031 - Security techniques — Guidelines for information and communication technology readiness for business continuity.

ITIL has defined some of these terms.

Recovery Time Objective

The Recovery Time Objective (RTO) is the targeted duration of time and a service level within which a business process must be restored after a disaster (or disruption) in order to avoid unacceptable consequences associated with a break in business continuity.

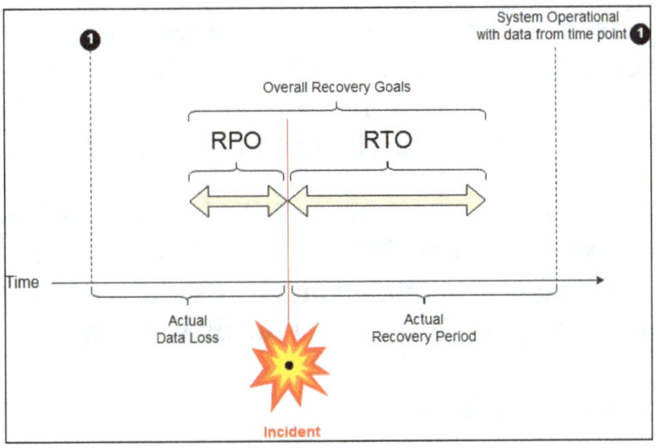

The terms RPO and RTO. In this example, the agreed values of RPO and RTO are *not* fulfilled.

In accepted business continuity planning methodology, the RTO is established during the Business Impact Analysis (BIA) by the owner of a process, including identifying options time frames for alternate or manual workarounds.

In a good deal of the literature on this subject, RTO is spoken of as a complement of Recovery Point Objective (RPO), with the two metrics describing the limits of acceptable or "tolerable" ITSC performance in terms of *time lost* (RTO) from normal business process functioning, and in terms of data lost or not backed up during that period of time (RPO) respectively.

Recovery Time Actual

A Forbes overview noted that it is *Recovery Time Actual* (RTA) which is "the critical metric for business continuity and disaster recovery."

RTA is established during exercises or actual events. The business continuity group times rehearsals (or actuals) and makes needed refinements.

Recovery Point Objective

A Recovery Point Objective (RPO) is defined by business continuity planning. It is the maximum targeted period in which data (transactions) might be lost from an IT service due to a major incident.

If RPO is measured in minutes (or even a few hours), then in practice, off-site mirrored backups must be continuously maintained; a daily off-site backup on tape will not suffice.

Relationship to Recovery Time Objective

Recovery that is not instantaneous will restore data/transactions over a period of time; the goal is to do so without incurring significant risks or significant losses.

RPO measures the maximum time period in which recent data might have been permanently lost in the event of a major incident; it is not a direct measure of the quantity of such loss. For instance if the BC plan is "restore up to last available backup", the RPO is the maximum interval between such backup that has been safely vaulted offsite.

Business impact analysis is used to determine RPO for each service; RPO is not determined by the existent backup regime. When any level of preparation of off-site data is required, the period during which data might be lost often starts near the time of the beginning of the work to prepare backups, not the time the backups are taken off-site.

Data Synchronization Points

Although a data synchronization point is a point in time, the timing for performing the physical backup must be included. One approach used is to halt processing of an update queue, while a disk-to-disk copy is made. The backup reflects the earlier time of that copy operation, not when the data is copied to tape or transmitted elsewhere.

How RTO and RPO Values affect Computer System Design

RTO and the RPO must be balanced, taking business risk into account, along with all the other major system design criteria.

RPO is tied to the times backups are sent offsite. Offsiting via synchronous copies to an offsite mirror allows for most unforeseen difficulty. Use of physical transportation for tapes (or other transportable media) comfortably covers some backup needs at a relatively low cost. Recovery can be enacted at a predetermined site. Shared offsite space and hardware completes the package needed.

For high volumes of high value transaction data, the hardware can be split across two or more sites; splitting across geographic areas adds resiliency.

Planning for disaster recovery and information technology (IT) developed in the mid- to late 1970s as computer center managers began to recognize the dependence of their organizations on their computer systems.

At that time, most systems were batch-oriented mainframes. Another offsite mainframe could be loaded from backup tapes pending recovery of the primary site; downtime was relatively less critical.

The disaster recovery industry developed to provide backup computer centers. One of the earliest such centers was located in Sri Lanka (Sungard Availability Services, 1978).

During the 1980s and 90s, as internal corporate timesharing, online data entry and real-time processing grew, more availability of IT systems was needed.

Regulatory agencies became involved even before the rapid growth of the Internet during the 2000s; objectives of 2, 3, 4 or 5 nines (99.999%) were often mandated, and high-availability solutions for hot-site facilities were sought.

IT Service Continuity is essential for many organizations in the implementation of Business Continuity Management (BCM) and Information Security Management (ICM) and as part of the implementation and operation information security management as well as business continuity management as specified in ISO/IEC 27001 and ISO 22301 respectively.

The rise of cloud computing since 2010 continues that trend: nowadays, it matters even less where computing services are physically served, just so long as the network itself is sufficiently reliable (a separate issue, and less of a concern since modern networks are highly resilient by design). 'Recovery as a Service' (RaaS) is one of the security features or benefits of cloud computing being promoted by the Cloud Security Alliance.

Classification of Disasters

Disasters can be the result of three broad categories of threats and hazards. The first category is natural hazards that include acts of nature such as floods, hurricanes, tornadoes, earthquakes, and epidemics. The second category is technological hazards that include accidents or the failures of systems and structures such as pipeline explosions, transportation accidents, utility disruptions, dam failures, and accidental hazardous material releases.

The third category is human-caused threats that include intentional acts such as active assailant attacks, chemical or biological attacks, cyber attacks against data or infrastructure, and sabotage. Preparedness measures for all categories and types of disasters fall into the five mission areas of prevention, protection, mitigation, response, and recovery.

Importance of Disaster Recovery Planning

Recent research supports the idea that implementing a more holistic pre-disaster planning approach is more cost-effective in the long run. Every $1 spent on hazard mitigation(such as a disaster recovery plan) saves society $4 in response and recovery costs.

2015 disaster recovery statistics suggest that downtime lasting for one hour can cost:

- Small companies as much as $8,000,
- Mid-size organizations $74,000,
- Large enterprises $700,000.

As IT systems have become increasingly critical to the smooth operation of a company, and arguably the economy as a whole, the importance of ensuring the continued operation of those systems, and their rapid recovery, has increased. For example, of companies that had a major loss of business data, 43% never reopen and 29% close within two years. As a result, preparation for continuation or recovery of systems needs to be taken very seriously. This involves a significant investment of time and money with the aim of ensuring minimal losses in the event of a disruptive event.

Control Measures

Control measures are steps or mechanisms that can reduce or eliminate various threats for organizations. Different types of measures can be included in disaster recovery plan (DRP).

Disaster recovery planning is a subset of a larger process known as business continuity planning and includes planning for resumption of applications, data, hardware, electronic communications (such as networking) and other IT infrastructure. A business continuity plan (BCP) includes planning for non-IT related aspects such as key personnel, facilities, crisis communication and reputation protection, and should refer to the disaster recovery plan (DRP) for IT related infrastructure recovery/continuity.

IT disaster recovery control measures can be classified into the following three types:

- Preventive measures: Controls aimed at preventing an event from occurring.
- Detective measures: Controls aimed at detecting or discovering unwanted events.

- Corrective measures: Controls aimed at correcting or restoring the system after a disaster or an event.

Good disaster recovery plan measures dictate that these three types of controls be documented and exercised regularly using so-called "DR tests".

Strategies

Prior to selecting a disaster recovery strategy, a disaster recovery planner first refers to their organization's business continuity plan which should indicate the key metrics of Recovery Point Objective and Recovery Time Objective. Metrics for business processes are then mapped to their systems and infrastructure.

Failure to properly plan can extend the disaster's impact. Once metrics have been mapped, the organization reviews the IT budget; RTO and RPO metrics must fit with the available budget. A cost-benefit analysis often dictates which disaster recovery measures are implemented.

Adding cloud-based backup to the benefits of local and offsite tape archiving, the *New York Times* wrote, "adds a layer of data protection."

Common strategies for data protection include:

- Backups made to tape and sent off-site at regular intervals.
- Backups made to disk on-site and automatically copied to off-site disk, or made directly to off-site disk.
- Replication of data to an off-site location, which overcomes the need to restore the data (only the systems then need to be restored or synchronized), often making use of storage area network (SAN) technology.
- Private Cloud solutions which replicate the management data (VMs, Templates and disks) into the storage domains which are part of the private cloud setup. These management data are configured as an xml representation called OVF (Open Virtualization Format), and can be restored once a disaster occurs.
- Hybrid Cloud solutions that replicate both on-site and to off-site data centers. These solutions provide the ability to instantly fail-over to local on-site hardware, but in the event of a physical disaster, servers can be brought up in the cloud data centers as well.
- The use of high availability systems which keep both the data and system replicated off-site, enabling continuous access to systems and data, even after a disaster (often associated with cloud storage).

In many cases, an organization may elect to use an outsourced disaster recovery provider to provide a stand-by site and systems rather than using their own remote facilities, increasingly via cloud computing.

In addition to preparing for the need to recover systems, organizations also implement precautionary measures with the objective of preventing a disaster in the first place. These may include:

- Local mirrors of systems and data and use of disk protection technology such as RAID.
- Surge protectors: to minimize the effect of power surges on delicate electronic equipment.
- Use of an uninterruptible power supply (UPS) and backup generator to keep systems going in the event of a power failure.
- Fire prevention/mitigation systems such as alarms and fire extinguishers.
- Anti-virus software and other security measures.

Control Measures

In disaster recovery planning it is important to be aware of the different types of control measures while designing the plan. Control measures, in and of themselves, are mechanisms or steps that can eliminate (or at least reduce) all of the different types of disaster threats that a business may face. There are different types of control measures that can be implemented into a disaster recovery plan. Disaster and Catastrophe offers a simple explanation on the different types of measures.

Disaster recovery planning is actually a subset of business continuity planning. A disaster recovery plan includes planning for getting applications back up and running, data, hardware, electronic communication methods (such as networking), and other vital parts of the IT infrastructure within the business. A business continuity plan involves planning for the non IT related factors and key aspects of a disaster such as facilities, personnel, crisis communication, and reputation protection. The business continuity plan should refer to the disaster recovery plan when dealing with the continuity or recovery of IT related infrastructure.

When related to IT recovery control measures, disaster recover control measures can be classified into three groups. The first is preventive measures. Preventive measures are the measures put into place with the intention of creating policies that aim to prevent an unfortunate event (or disaster) from actually occurring. The second type of control measures are detective measures. Detective measures are the measures put into place with the intention of creating policies that spell out the process for detecting or discovering unwanted events. These measures are put into place to find an event that is in place but hasn't occurred yet and the policies spell out how to handle them before they become a disaster. The third type of control measure are corrective measures. Corrective measures are the measures that are put into place with the intention of creating policies and procedures that will handle correcting or restoring the system or network after a disaster has occurred. Corrective measures are aimed at getting the business back up and running as soon as possible.

There are many different ways to implement corrective measures into a plan. The best way to go about developing a plan and making sure that all of the necessary corrective measures are in place is to consult with professionals (like the ones found at Virtacore). This way no measures are forgotten, all areas are covered, and in the event of a disaster that con not be prevented your company is safe and the damage is minimal. Making sure efficient corrective measures are in place will save the company both money and time in the long run.

Practices in Disaster Recovery

When a disaster strikes, the normal operations of the enterprise are suspended and replaced with operations spelled out in the disaster recovery plan. Figure depicts the cycle of stages that lead through a disaster back to a state of normalcy.

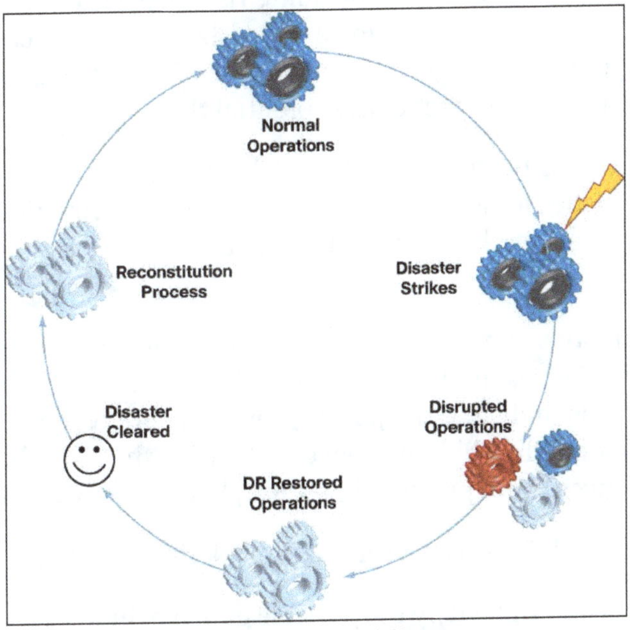

Enterprise operations cycle of disaster recovery.

It takes the enterprise some time to assess the exact effects of the disaster. Only when these are assessed and the affected systems are identified can a recovery process begin. The disaster recovery system cannot replace the normal working system forever, but only supports it for a short period of time. At the earliest possible time, the disaster recovery process must be decommissioned and the business should return to normalcy.

The disaster recovery plan does not stop at defining the resources or processes that need to be in place to recover from a disaster. The plan should also define how to restore operations to a normal state once the disaster's effects are mitigated. Finally, ongoing procedures for testing and improving the effectiveness of the disaster recovery system are part of a good disaster recovery plan.

In summary, the disaster recovery plan should (1) identify and classify the threats/risks that may lead to disasters, (2) define the resources and processes that ensure business continuity during the disaster, and (3) define the reconstitution mechanism to get the business back to normal from the disaster recovery state, after the effects of the disaster are mitigated. An effective disaster recovery plan plays its role in all stages of the operations as depicted above, and it is continuously improved by disaster recovery mock drills and feedback capture processes.

Disaster Recovery Planning

Identification and Analysis of Disaster Risks/Threats

The first step in planning recovery from unexpected disasters is to identify the threats or risks that can bring about disasters by doing risk analysis covering threats to business continuity. Risk analysis (sometimes called business impact analysis) involves evaluating existing physical and environmental security and control systems, and assessing their adequacy with respect to the potential threats.

The risk analysis process begins with a list of the essential functions of the business. This list will set priorities for addressing the risks. Essential functions are those whose interruption would considerably disrupt the operations of the business and may result in financial loss.

These essential functions should be prioritized based on their relative importance to business operations. For example, in the case of a telecom service provider, though both billing operations and CRM/helpdesk operations are essential functions, CRM/helpdesk is less essential than billing. Hence, mitigating the risks that affect billing operations should be given more priority than CRM/helpdesk operations.

While evaluating the risks, it is also useful to consider the attributes of a risk.

The scope of a risk is determined by the possible damage, in terms of downtime or cost of lost opportunities. In evaluating a risk, it is essential to keep in mind the options around that risk, such as time of the day or day of the week, that can affect its scope. For example, spilling several gallons of toxic liquid across an assembly line area during working hours is a different situation than the same spill at night or during the weekend. While the time taken and cost to clean up the area are the same in both cases, the first case may require shutting down the assembly line area, which adds downtime cost to this event.

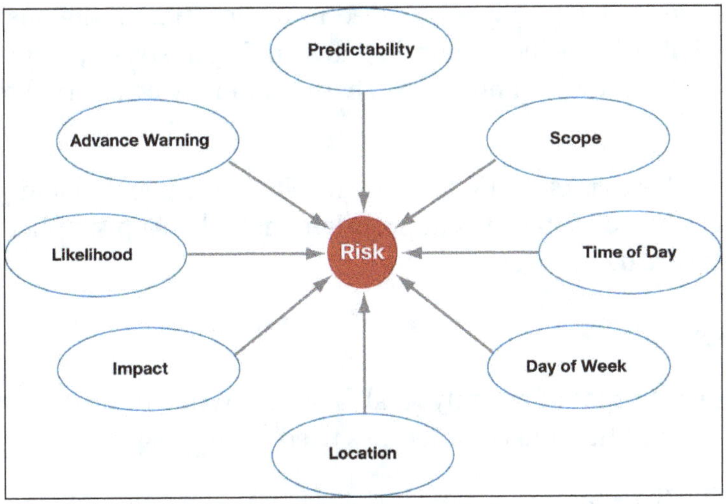

Risk attributes.

The magnitude of a risk may be different considering the affected component, its location, and the time of occurrence. The effects of a disaster that strikes the entire enterprise are different from the effects of a disaster affecting a specific area, office, or utility within the company.

Classification of Risks based on Relative Weights

When evaluating risks, it is recommended to categorize them into different classes to accurately prioritize them. In general, risks can be classified in the following five categories.

External Risks

External risks are those that cannot be associated with a failure within the enterprise. They are very significant in that they are not directly under the control of the organization that faces the damages. External risks can be split into four subcategories:

- Natural: These disasters are on top of the list in every disaster recovery plan. Typically they damage a large geographical area. To mitigate the risk of disruption of business operations, a recovery solution should involve disaster recovery facilities in a location away from the affected area. Nowadays most of the meteorological threats can be forecasted, hence the chances to mitigate effects of some natural disasters are considerable. Nevertheless is important to consider documenting the scope of these natural risks in as much detail as possible.
- Human caused: These disasters include acts of terrorism, sabotage, virus attacks, operations mistakes, crimes, and so on. These also include the risks resulting from manmade structures. These may be caused by both internal and external persons.

- Civil: These risks typically are related to the location of the business facilities. Typical civil risks include labor disputes ending in strikes, communal riots, local political instability, and so on. These again may be internal to the company or external.

- Supplier: These risks are tied to the capacity of suppliers to maintain their level of services in a disaster. It is appropriate that a backup supplier pool be maintained in case of emergency.

Facility Risks

Facility risks are risks that affect only local facilities. While evaluating these risks, the following essential utilities and commodities need to be considered.

- Electricity: To analyze the power outage risk, it is important to study the frequency of power outage and the duration of each outage. It is also useful to determine how many powers feeds operate within the facility and if necessary make the power system redundant.

- Telephones: Telephones are a particularly crucial service during a disaster. A key factor in evaluating risks associated with telephone systems is to study the telephone architecture and determine if any additional infrastructure is required to mitigate the risk of losing the entire telecommunication service during a disaster.

- Water: There are certain disaster scenarios where water outages must be considered very seriously, for instance the impact of a water cutoff on computer cooling systems.

- Climate Control: Losing the air conditioning or heating system may produce different risks that change with the seasons.

- Fire: Many factors affect the risk of fire, for instance the facility's location, its materials, neighboring businesses and structures, and its distance from fire stations. All of these and more must be considered during risk evaluation.

- Structural: Structural risks may be related to design flaws, defective material, or poor-quality construction or repairs.

- Physical Security: Security risks have gained attention in recent years, and nowadays security is a mandatory 24-hour measure to protect each and every asset of the company from both outsiders and employees. Different secure access and authorization procedures, manual as well as automated ones, are enforced in enterprises. Factors such as workplace violence, bomb threats, trespassing, sabotage, and intellectual property loss are also considered during the security risk analysis.

Data Systems Risks

Data systems risks are those related to the use of shared infrastructure, such as networks, file servers, and software applications that could impact multiple departments. A key objective in analyzing these risks is to identify all single points of failure within the data systems architecture.

Data systems risks can also be due to inappropriate operation processes. Operations that have run for a long period of time on obsolete hardware or software are a major risk given the lack of spares or support. Recovery from this type of failure may be lengthy and expensive due to the need to replace or update software and equipment and retrain personnel.

Data systems risks may be evaluated within the following subcategories:

- Data communication network,
- Telecommunication systems and network,
- Shared servers,
- Virus,
- Data backup/storage systems,
- Software applications and bugs.

Departmental Risks

Departmental risks are the failures within specific departments. These would be events such as a fire within an area where flammable liquids are stored, or a missing door key preventing a specific operation.

An effective departmental risk assessment needs to consider all the critical functions within that department, key operating equipment, and vital records whose absence or loss will compromise operations. Unavailability of skilled personnel also can be a risk. The department should have necessary plans to have skilled backup personnel in place.

Desk-level Risks

Desk-level risks are all the risks that can happen that would limit or stop the day-to-day personal work of an individual employee. The assessment at this layer may feel a little like an exercise in paranoia. Every process and tool that makes up the personal job must be examined carefully and accounted as essential.

Building the Risk Assessment

Once the evaluation of the major risk categories is completed, it is time to score and sort all of them, category by category, in terms of their likelihood and impact. The scoring process can be approached by preparing a score sheet, as shown in table, that has the following keys:

- Groups are the subcategories of the main risk category.
- Risks are the individual risks under each group that can affect the business.
- Likelihood is estimated on a scale from 0 to 10, with 0 being not probable and 10 highly probable. The likelihood that something happens should be considered in a long plan period, such as 5 years.
- Impact is estimated on a scale from 0 to 10, with 0 being no impact and 10 being an impact that threatens the company's existence. Impact is highly sensitive to time of day and day of the week.
- Restoration Time is estimated on a scale from 1 to 10. A higher value would mean longer restoration time hence the priority of having a Disaster Recovery mechanism for this risk is higher.

Table: Risk Assessment Form

Risk Assessment Form					
External risks					
Date:		Likelihood	Impact	Restoration Time	Score
Grouping	Risk	0 – 10	0 – 10	1 – 10	
Natural disasters					
	Earthquake	1	9	10	90
	Tornado	0	0	10	0
	Severe thunderstorm				0
	Hail	8	3	9	216
	Snow/ice/blizzard	9	5	8	360
Human caused risks					
	Sabotage or act of terror				
	Bridge collapse				
	Water leakage in facility				
Civil issues					
	Riot				
	Labor stoppage and picketing				
Suppliers					
	Power supplier				
	Transportation vendor				

Looking at the above example, multiplying the likelihood time, impact time, and restoration time yields a rough risk analysis score. A zero value within one of the two columns makes the total risk score a zero. Sorting the table in descending order will put the biggest risks to the top, and these are the risks that deserve more attention.

Determining the effects of Disasters

Once the disaster risks have been assessed and the decision has been made to cover the most critical risks, the next step is to determine and list the likely effects of each of the disasters. These specific effects are what will need to be covered by the disaster recovery process.

Simple "one cause multiple effects" diagrams can be used as tools for specifying the effects of each of the disasters.

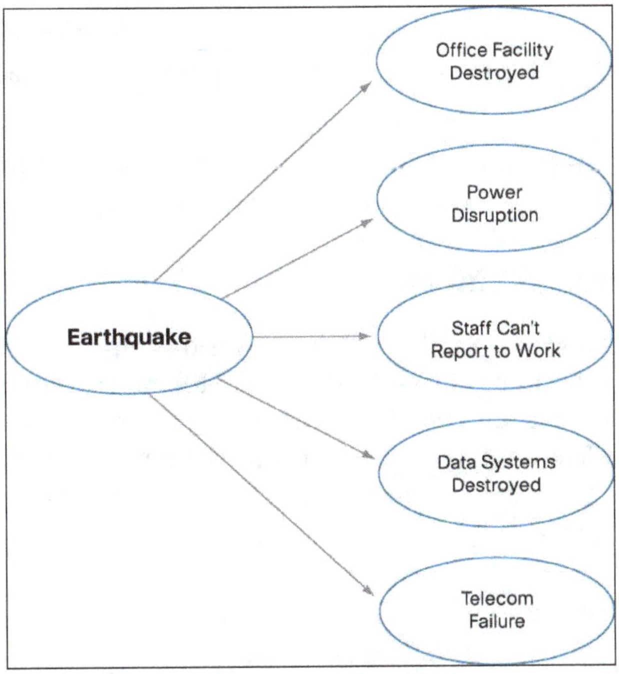

Disaster effects diagram.

Note that multiple causes can produce the same effects, and in some cases the effects themselves may be the causes of some other effects.

List of Disaster affected Entities

The intention of this exercise is to produce a list of entities affected by failure due to disasters, which need to be addressed by the disaster recovery plan. In figure, the entities that fail due to the earthquake disaster are office facility, power system, operations staff, data systems, and telephone system. Table provides a sample mapping of the cause, effects, and affected entities.

Table: Determination of disaster affected entities.

Risk (Disaster)	Effect of Disaster	Disaster Affected Entity
Earthquake	Office space destroyed	Office space
	Operators cannot report to work	Office staff
	Power disruption	Power
	Data systems destroyed	Data sysatems
	Desktops destroyed	Desktops and workstations
	Telecom failure	Telephone instruments and network
Power supply cut	Power disruption	Power
	Data systems powered off	Data systems
	Desktops powered off	Desktops/workstations
	Data network down	Network devices and links
	Telecom failure	Telephone instruments and network

It may be noticed that two or more disasters may affect the same entities, and it can be determined which entities are affected most often. The entities with the most appearances in the table have a greater tendency of failure occurrence.

Downtime Tolerance Limits

Once the list of entities that possibly fail due to various types of disasters is prepared, the next step is to determine what is the downtime tolerance limit for each of the entities. This information becomes crucial for preparing the recovery sequence in the disaster recovery plan. The entities with less downtime tolerance limit should be assigned higher priorities for recovery. One metric for evaluating the downtime tolerance limit is the cost of downtime.

Cost of Downtime

The cost of downtime is the main key to calculate the investment needed in a disaster recovery plan. Downtime costs can be divided into tangible and intangible costs.

Tangible costs are those costs that are a consequence of a business interruption, generating loss of revenue and productivity. Intangible costs include lost opportunities when customers would approach competitors, loss of reputation, and similar factors.

Interdependencies

How the disaster affected entities depend upon each other is crucial information for preparing the recovery sequence in the disaster recovery plan. For example, having the data systems restored has a dependency on the restoration of power.

Evaluation of Disaster Recovery Mechanisms

Once the list of affected entities is prepared and each entity's business criticality and failure tendency is assessed, it is time to analyze various recovery methods available for each entity and determine the best suitable recovery method for each. This step defines the resources employed in recovery and the process of recovery. Some of the typical entities are data systems, power, data network, and telephone systems. For each of these there are one or more recovery mechanisms in practice in the industry.

In the case of data systems, for example, the recovery mechanism usually involves having the critical data systems replicated somewhere else in the network and putting them online with the latest backed up data available. For less critical data systems, there may be an option to have spare server hardware, and if required these servers could be configured with the required application. Depending on the data system, there may be options of autorecovery or manual recovery, and the cost and recovery time factors of each mechanism vary.

In the case of power, options such as multiple power suppliers or having alternate sources of power such as diesel generators may be suitable. In certain cases, new mechanisms may need to be devised.

Considering multiple options and variations of disaster recovery mechanisms available, it is necessary to carefully evaluate the best suitable recovery mechanism for an affected entity in a particular organization. The main factors that need to be considered are:

- Cost of deployment, maintenance, and operation.
- Recovery time.
- Ease of recovery activation and operation.

Disaster Recovery Committee

Disaster recovery operations and procedures should be governed by a central committee. This committee should have representation from all the different company agencies with a role in the disaster recovery process, typically management, finance, IT (multiple technology leads), electrical department, security department, human resources, vendor management, and so on.

The Disaster Recovery Committee creates the disaster recovery plan and maintains it. During a disaster, this committee ensures that there is proper coordination between different agencies and that the recovery processes are executed successfully and in proper sequence.

The Disaster Recovery Committee should be authorized and responsible for:

- Creating and maintaining the disaster recovery plan.

- Detecting and announcing disaster events within the company.
- Activating the disaster recovery plan.
- Executing the disaster recovery plan.
- Monitoring the disaster situation continuously and returning operations to normal at the earliest feasible time.
- Restoring normal operations and shutting down disaster recovery operations.
- Continuously improving the disaster recovery plan by conducting periodic mock trials and incorporating lessons learned into the plan after an actual disaster.

The roles, responsibilities, and reporting hierarchy of different committee members should be clearly defined both during normal operations and in the case of a disaster emergency. Backup members should also be designated in case of the primary member's unavailability.

Note that not all the members of the Disaster Recovery Committee may actively participate in the actual disaster recovery. But several key members of the committee, such as the operations manager, operations coordinator, and the respective operations team leads, will always actively participate.

Disaster Recovery Phases

Disaster recovery happens in the following sequential phases:

- Activation Phase: In this phase, the disaster effects are assessed and announced.
- Execution Phase: In this phase, the actual procedures to recover each of the disaster affected entities are executed. Business operations are restored on the recovery system.
- Reconstitution Phase: In this phase the original system is restored and execution phase procedures are stopped.

Activation Phase

A disruption or emergency may happen with or without notice. A hurricane affecting a specific geographic area, or a virus spread expected on a certain date are examples of disasters with advance notice. However, there may be no warning of the burst of a water pipe or a human criminal act.

Quick and precise detection of a disaster event and having an appropriate communication plan are the key for reducing the effects of the incoming emergency; in some cases it may give enough time to allow system personnel to implement actions gracefully, thus reducing the impact of the disaster.

The Disaster Recovery Committee is responsible for launching the activation phase. It should be well informed about the geographical, political, social, and environmental events that may pose threats to the company's business operations. It should have trusted information sources in the different agencies to forestall false alarms or overreactions to hoaxes.

The activation phase involves:

- Notification procedures.
- Damage assessment.
- Disaster recovery activation planning.

Notification Procedures

The notification procedure defines the primary measures taken as soon as a disruption or emergency has been detected or definitely predicted. At the end of this phase, recovery staff will be ready to execute contingency actions to restore system functions on a temporary basis. Procedures should contain the process to alert recovery personnel during business and nonbusiness hours. After the disaster detection, a notification should be sent to the damage assessment team, so that they can assess the real damage occurred and implement subsequent actions.

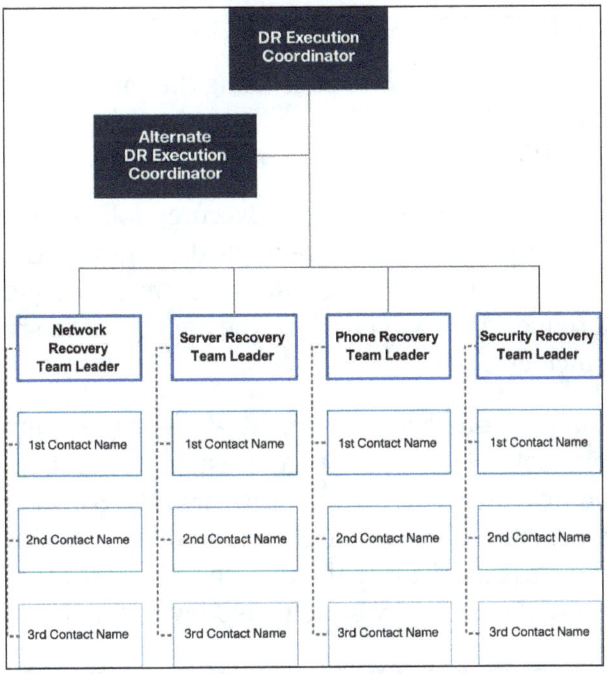

All tree chart.

Notification can take place by telephone, pager, e-mail, or cell phone. A notification policy must describe procedures to be followed when specific personnel cannot be contacted. Notification procedures should be documented clearly in the contingency plan.

A general notification technique is a call tree. The call tree should document primary and alternate contact methods and should include procedures to be followed if an individual cannot be contacted.

Staff to be alerted should be unmistakably identified in the contact list in the plan. This list should classify personnel by their role, name, and contact information (home, work, and pager numbers, e-mail addresses, and home addresses). If disrupted systems have interconnection with external organizations, a point of contact should be identified in those organizations.

Notification information may contain the following:

- Nature of the emergency that has occurred or is imminent.
- Loss of life or injuries.
- Damage estimates.
- Response and recovery details.
- Where and when to assemble for briefing or further response instructions.
- Instructions to prepare for relocation for estimated time period.
- Instructions to complete notifications using the call tree (if applicable).

Damage Assessment

To establish how the contingency plan will be executed following a service disruption, it is crucial to evaluate the nature and degree of the damage to the system. This damage evaluation should be done as quickly as conditions permit, with personnel safety given highest priority. Consequently, when possible, the damage assessment team is the first team notified of the incident.

It is worthwhile to prepare damage assessment guidelines for investigating different types of major alarms that may progress to a disaster. An example might be a sudden power outage noticed in a data center facility that has a UPS backup. The investigation may determine whether the power can be restored before the UPS system runs out of battery power, in which case activating the disaster recovery plan is not necessary, or otherwise, in which case the plan may be activated immediately.

Damage assessment procedures vary with each particular emergency; nevertheless, the following may be considered in general:

- Origin of the emergency or disruption.

- Potential for additional disruptions or damage.
- Area affected by the emergency.
- Status of physical infrastructure.
- Inventory and functional status of the most important equipment.
- Type of damage to equipment.
- Items to be replaced.
- Estimated time to restore normal services if disaster procedures were not in place.

Activation Planning

While it is beneficial to detect a disaster at its earliest stage, putting a disaster recovery process into action for a false alarm may stall normal business operations and result in undue costs. Hence it is very important that disaster recovery be activated only when a thorough damage assessment has been conducted.

The disaster recovery plan should have one or more criteria for activation, which become the primary input for evaluating whether the plan should be activated for each affected system. Also, it should be determined whether activating disaster response will bring systems back on line faster than standard procedures.

Depending on the extent of the damage from the disaster, the entire Disaster Recovery Committee or a part of the committee may do the disaster activation planning. The outcome of this planning, at a minimum, should be:

- List of systems and services that need to be restored.
- Their interdependencies and sequence of restoration.
- Time estimations for each restoration (documented in the plan).
- Instructions for reporting failures to the team leads.
- Plan for communication between teams.

Once the disaster activation is planned, the appropriate team leads will notify staff and start their respective activities in sequence as they have been instructed.

Execution Phase

Recovery operations start just after the disaster recovery plan has been activated, appropriate operations staff have been notified, and appropriate teams have been mobilized.

The activities of this phase focus on bringing up the disaster recovery system. Depending on the recovery strategies defined in the plan, these functions could include temporary manual processing, recovery and operation on an alternate system, or relocation and recovery at an alternate site.

Sequence of Recovery Activities

The recovery procedure reflects priorities previously analyzed during the activation planning phase. For instance, if a server room has been recovered after a disruption, the most critical servers should be restored before other, less critical servers. The procedures should also include instructions to coordinate with other teams when certain situations occur, such as:

- An action is not accomplished within the estimated time frame.
- A key step has been completed.
- Items must be procured.

If a system must be recovered at a different location, specific items related to that service need to be transferred or obtained. Recovery procedures should delegate a team to manage shipment of equipment, data, and vital records. Procedures should explain requirements to package, transport, and purchase materials required to recover the system.

Recovery Procedures

The disaster recovery plan should provide detailed procedures to restore the system or system components. Procedures for IT service damage should address specific actions such as:

- Get authorization to access damaged premises or geographic area.
- Notify users associated with the system.
- Obtain required office supplies and work space.
- Obtain and install required hardware components.
- Obtain and load backup media.
- Restore critical operating systems and application software.
- Restore system data.
- Test system functionality including security controls.
- Connect system to network or other external systems.

To avoid confusion in an emergency situation, the recovery procedures should be documented in a simple step-by-step format, without assuming or omitting any procedural steps.

Reconstitution Phase

In the reconstitution phase, operations are transferred back to the original facility once it is free from the disaster aftereffects, and execution-phase activities are subsequently shut down. If the original system or facility is unrecoverable, this phase also involves rebuilding. Hence the reconstitution phase may last for a few days to few weeks or even months, depending on the severity of destruction and the site's fitness for restoration. As soon as the facility, whether repaired or replaced, is able to support its normal operations, the services may be moved back. The execution team should continue to be engaged until the restoration and testing are complete.

The following major activities occur in this phase:

- Continuously monitor the site or facility's fitness for reoccupation.
- Verify that the site is free from aftereffects of the disaster and that there are no further threats.
- Ensure that all needed infrastructure services, such as power, water, telecommunications, security, environmental controls, office equipment, and supplies, are operational.
- Install system hardware, software, and firmware.
- Establish connectivity between internal and external systems.
- Test system operations to ensure full functionality.
- Shut down the contingency system.
- Terminate contingency operations.
- Secure, remove, and relocate all sensitive materials at the contingency site.
- Arrange for operations staff to return to the original facility.

The Disaster Recovery Plan Document

The outcome of the disaster recovery planning process is the disaster recovery plan document. During an emergency, this document will be the primary source of information for disaster recovery procedures.

Document Contents

The disaster recovery plan document is the only reliable source of information for the disaster recovery during an emergency. It should be very easily readable, with simple and detailed instructions. Following are some of the contents that need to be in this document:

- Document Information: The document should include information such as the authors/owners with their contact details, revision history and other document details (name, location, version), references, and the audience of the document. In the document revision history, it is good to have a brief description of the changes made in each version. A table of contents is a must for quick reference, and it is highly recommended that the sections be numbered to the lowest possible level for easy reference purpose. It is also good to give an appropriate confidential status for the document as it contains sensitive information.
- Purpose: The purpose of the document must be clearly stated in the introduction, defining the objectives the plan intends to achieve.
- Scope: The scope of the plan defines the circumstances under which the plan is invoked and the length of time the procedures defined in the document are in effect. The different failure conditions that lead to invoking the plan should be clearly listed. For example, a system being down for couple of hours may not result in invoking the plan, but a daylong outage may suffice. Similarly, the conditions at the failed system/facility that warrant the reconstitution phase should also be clearly stated.
- Assumptions: Any conditions the plan assumes to be present for success should be clearly stated. This may involve listing the dependencies of the plan as well. For example, a certain number of trained personnel may be assumed to be available at the disaster recovery facility. Wherever possible, these dependencies must be accompanied with the appropriate contact details.
- Exclusions: Any related disaster activities that the plan does not cover should be stated and any known references mentioned here. For example, the plan may exclude the dependent power restoration plan, referring instead to the appropriate document and the department contact details. Such information will be useful during the disaster recovery.
- System Description: The description of the disaster recovery system should be simple to understand with appropriate figures, workflow charts, and so on. If necessary the descriptions may reference appendices that give more detail. The functions that need to be revived need to be clearly mentioned.
- Roles and Responsibilities: The roles of the managerial and technical staff and their responsibilities during the activation, execution, and reconstitution phases should be clearly listed. An organization structure diagram showing the reporting relationships is beneficial. Key roles should have primary and alternate personnel assigned.

- Contact Details: Full contact information should be included for all the managerial and technical staff involved in the planning, activation, execution, and reconstitution phases. Contact details both during normal situations and emergency situations should be mentioned. This information is recommended to be added as an appendix to the disaster recovery plan document.
- Activation Procedures: The procedures for notification, damage assessment, and activation planning should be outlined. Any topic that needs to be covered in great detail may be added as an appendix.
- Execution Procedures: The recovery procedure for each of the components the plan covers should be explained step by step in detail. When there are parallel threads of tasks, it is beneficial to have a flow chart diagram to visualize the dependencies of the tasks. The success and failure criteria of each procedure also should be mentioned as well as instructions on further actions in case of both success and failure.
- Reconstitution Procedures: Similar procedures for the reconstitution of the components should be explained in detail. The success and failure criteria and instructions for further actions in case of success and failure should be given.

Document Maintenance

The disaster recovery plan document needs to be kept up to date with the current organization environment. A plan that is not updated and tested is as bad as not having a plan at all because during emergencies, the document may be misleading. The following are recommended for maintenance of the plan documentation.

- Periodic Mock Drills: The disaster recovery plan should be tested from time to time using scheduled mock drills. A drill usually will not affect active operations; however, if it is known that operations will be affected, the drill should be carefully scheduled such that the effect is minimal and is done during a permissible window. These activities should be regarded similarly to regular equipment maintenance activities that require operations downtime. The experience of the mock drill should be updated into the disaster recovery plan document.
- Experience Capture: The best testing the document will undergo is when an actual disaster happens, and the lessons learned during the disaster recovery are valuable for improving the plan. Hence the Disaster Recovery Committee should ensure that the experience gets captured as lessons learned and the document gets updated accordingly.
- Periodic Updates: Technologies, systems, and facilities that the plan covers may change over time. It is important that the disaster recovery plan document reflect the current information about the components it covers. For this purpose, the Disaster Recovery Committee should ensure that the document is audited

periodically (say once every quarter) against the present components in the organization. Another way to achieve this is to ensure that the committee is notified of any change that happens to any system/component in the organization so that the committee may update the document accordingly.

Mitigation

Disaster Mitigation is the cornerstone of emergency management. It's the ongoing effort to lessen the impact disasters have on people and property. Mitigation involves keeping homes away from floodplains, engineering bridges to withstand earthquakes, creating and enforcing effective building codes to protect property from hurricanes, and more.

Mitigation is defined as "sustained action that reduces or eliminates long-term risk to people and property from natural hazards and their effects." It describes the ongoing effort at the federal, state, local and individual levels to lessen the impact of disasters upon our families, homes, communities and economy.

How Disaster Mitigation can Help You

Through the application of mitigation technologies and practices, our society can ensure that fewer Americans and their communities become victims of natural disasters. For example, mitigation measures can be applied to strengthen your home, so that your family and belongings are better protected from floods, earthquakes, hurricanes, and other natural hazards. They can be utilized to help business and industry to avoid damage to their facilities and remain operational in the face of catastrophe. Mitigation technologies can be used to strengthen hospitals, fire stations, and other critical service facilities so that they can remain operational or reopen more quickly after an event. In addition, mitigation measures can help reduce disaster losses and suffering so that there is less demand for money and resources in the aftermath.

In practice, mitigation can take many forms. It can involve actions such as:

- Promoting sound land use planning based on known hazards.
- Buying flood insurance to protect your belongings.
- Relocating or elevating structures out of the floodplains.
- Securing shelves and water heaters to nearby walls.
- Having hurricane straps installed to more securely attach a structure's roof to its walls and foundation.
- Developing, adopting, and enforcing effective building codes and standards.
- Engineering roads and bridges to withstand earthquakes.

- Using fire-retardant materials in new construction.
- Developing and implementing a plan in your business or community to reduce your susceptibility to hazards.

Mitigation is a type of long-term, pre-disaster planning which involves sustained expenditures on structural and non-structural efforts to reduce or eliminate future risks. Mitigation plans and activities are, in practice, usually medium to long term, and mitigation is the cornerstone of emergency management since it is an example where thinking ahead pays off in the long run. Terminologically, mitigation is related to two other concepts of long-term planning: reconstruction and preparedness. Reconstruction means repair or rebuilding, and preparedness means getting ready or practicing to respond. Mitigation means to lessen the effects or take action toward the building and putting together of certain structures and plans so that the impact of any future disaster will be ameliorated, or eliminated, if possible. Amelioration means to change things for the better, and impact can be understood as the consequences, or the likelihood of something happening in the first place, if the latter is theoretically possible.

Some simple examples of mitigation activities that an emergency manager might do include: promoting flood insurance, urging the structural redesign of buildings, raising or moving homes from flood zones, or just making sure there are appropriate building codes within certain communities. Mitigation planning involves an assessment of the threats facing a community, such as the likelihood of a terrorist attack, and an assessment of possible targets. Terrorist mitigation is a somewhat controversial phrase which implies special plans and practices for terrorism need to supplement an all-hazards approach. Mitigation planning is an ongoing process, with continual reassessments as necessary to ensure proper preparedness. Some experts argue that there is such as thing as post-disaster mitigation, and that pre-disaster mitigation ought to be called prevention. The usual division of mitigation into two categories - (1) structural and (2) non-structural - is intended to denote the importance of integrated planning in mitigation; that is, the kind of planning which efficiently balances a combination of engineering solutions (like moving homes) with political solutions (like changing the zoning abatements for a community). Some solutions only have a short window of opportunity to capitalize on public and political support. Non-structural solutions are often brought in when engineering solutions have become very costly and have not resulted in a substantial reduction in losses. Evacuation planning is sometimes considered a type of non-structural mitigation.

Aims of Disaster Mitigation

The specific aims of disaster mitigation are as follows:

- Creating awareness of risk at community level.
- To promote local actions through community participation to reduce such risks.

- Induction of mitigation measures in the national and regional development plans, land use planning proposals and in project design and appraisal in disaster prone areas.
- To assist decision makers (politicians and administrators) to understand the nature and extent of various risks faced by the communities in their area of responsibility, and to assess the economic effects of natural disasters on agriculture, commerce and industry.
- To demonstrate ways and means to reduce those risks within the limits of national, regional or socio-economic conditions through proper decision making and planning.
- To introduce effective measures to implement disaster mitigation plans at the different levels of public administration based on risk assessment and vulnerability analysis.

Mitigation Actions

A mitigation action is a specific action, project, activity, or process taken to reduce or eliminate long-term risk to people and property from hazards and their impacts. Implementing mitigation actions helps achieve the plan's mission and goals. The actions to reduce vulnerability to threats and hazards form the core of the plan and are a key outcome of the planning process.

Types of Mitigation Actions

The primary types of mitigation actions to reduce long-term vulnerability are:

- Local plans and regulations.
- Structural projects.
- Natural systems protection.
- Education programs.
- Preparedness and response actions.

Local Plans and Regulations

Local land use or comprehensive plans embody the goals, values and aspirations of the community, as expressed through a process of community engagement. The plan should identify current development patterns and trends as well as areas where future development should and should not occur. The plan should include policies and ordinances that steer development away from hazard-prone areas, such as floodplains, to avoid putting people and property at risk. In some cases, local plans can work at cross-purposes. For example, a capital improvement plan may call for extending water and sewer lines to an area that is vulnerable to natural hazards. Emergency managers, planners and others in a community should coordinate in preparing plans to ensure consistency across plans; that is, consistent goals, policies, and strategies.

Local ordinances and review processes influence the way land and buildings are developed and built. Examples include:

- Comprehensive plans.
- Land use ordinances.
- Subdivision regulations.
- Development review.
- Building codes and enforcement.
- NFIP Community Rating System.
- Capital improvement programs.
- Open space preservation.
- Stormwater management regulations and master plans.

Plans, ordinances, policies and regulations should be mutually reinforcing. All should leave to the development of a more sustainable, resilient community.

Structure and Infrastructure Projects

These actions involve modifying existing structures and infrastructure to protect them from a hazard or remove them from a hazard area. This could apply to public or private structures as well as critical facilities and infrastructure. This type of action also involves projects to construct manmade structures to reduce the impact of hazards. Many of these types of actions are projects eligible for funding through the FEMA Hazard Mitigation Assistance program. *Task 9 – Create a Safe and Resilient Community* provides more information on these programs. Examples include:

- Acquisitions and elevations of structures in flood prone areas.
- Utility undergrounding.
- Structural retrofits.
- Floodwalls and retaining walls.
- Detention and retention structures.

- Culverts.
- Safe rooms.

Natural Systems Protection

These are actions that minimize damage and losses and also preserve or restore the functions of natural systems. Examples include:

- Sediment and erosion control.
- Stream corridor restoration.
- Forest management.
- Conservation easements.
- Wetland restoration and preservation.

Education and Awareness Programs

These are actions to inform and educate citizens, elected officials, and property owners about hazards and potential ways to mitigate them. These actions may also include participation in national programs, such as StormReady1 or Firewise2 Communities. Although this type of mitigation reduces risk less directly than structural projects or regulation, it is an important foundation. A greater understanding and awareness of hazards and risk among local officials, stakeholders, and the public is more likely to lead to direct actions. Examples include:

- Radio or television spots.
- Websites with maps and information.
- Real estate disclosure.
- Presentations to school groups or neighborhood organizations.
- Mailings to residents in hazard-prone areas.
- StormReady.
- Firewise Communities.

Preparedness and Response Actions

Mitigation actions reduce or eliminate long-term risk and are different from actions taken to prepare for or respond to hazard events. Mitigation activities lessen or eliminate the need for preparedness or response resources in the future. When analyzing risks and identifying mitigation actions, the planning team may also identify emergency response or operational preparedness actions. Examples include:

- Creating mutual aid agreements with neighboring communities to meet emergency response needs.
- Purchasing radio communications equipment for the Fire Department.
- Developing procedures for notifying citizens of available shelter locations during and following an event.

For some hazards, such as tornadoes, including preparedness actions in the mitigation plan may be necessary and practical. The mitigation plan may be the best place for your community to capture and justify the need for these actions. However, these will not take the place of or meet the federal mitigation planning requirements for identifying mitigation actions. It is important that the planning team understands the difference and can distinguish between mitigation and other emergency management activities.

Common Mitigation Actions

- Enforcement of building codes, floodplain management codes and environmental regulations.
- Public safety measures such as continual maintenance of roadways, culverts and dams.
- Acquisition of relocation of structures, such as purchasing buildings located in a floodplain.
- Acquisition of undeveloped hazard prone lands to ensure no future construction occurs there.
- Retrofitting of structures and design of new construction, such as elevating a home or building.
- Protecting critical facilities and infrastructure from future hazard events.
- Planning for hazard mitigation, emergency operations, disaster recovery, and continuity of operations.
- Development and distribution of outreach materials related to hazard mitigation.
- Deployment of warning systems to alert and notify the public.

During the planning process, the City of Los Angeles Steering Committee is actively engaging community members and stakeholders in the planning process as part of a whole community approach in hazard mitigation and disaster planning.

Hazard Mitigation Plan Revision Process

This hazard mitigation planning process has six steps:

- Organize Resources & Build the Planning Team: Relevant studies, plans, and reports are collected along with communications resources that allow the public to be involved throughout the planning process. A planning team is assembled consisting of municipal representatives, and local and regional stakeholders.
- Develop the Plan's Risk Assessment: The risk assessment includes the identification of the location and geographic extent of natural and human-caused hazards that can affect the City. The hazard impacts and future probability of occurrence is also determined. Scientific and historical evidence of past events is collected and evaluated. All of these factors, along with the information on damage and losses sustained by the City, enables the hazards to be ranked from highest threat to lowest threat.
- Assess Capabilities: Local capabilities through emergency management, the National Flood Insurance Program, planning and regulatory authorities, administrative, technical, financial, and political capacities are assessed for the plan revision.
- Develop the Mitigation Strategy: Goals, objectives, and past mitigation actions are evaluated and revised as needed by the planning team. The planning team will also define appropriate new mitigation techniques, and prioritize mitigation actions and projects in the revised mitigation strategy.
- Determine Plan Maintenance Process: The HMP is a living document that must be regularly reviewed, updated, and maintained. A schedule is prepared to include responsible departments involved with monitoring, evaluating, and updating the plan during its five-year cycle. A process for integrating the updated Mitigation Strategy into existing plans and reports should be outlined and a plan for continued public outreach and participation must also be developed.
- Obtain Mitigation Plan Approval and Adoption: The draft plan is made available for public comment then submitted to the State of California Office of Emergency Services (Cal OES) and FEMA for review and approval. Once a Plan has been determined to meet all state and federal requirements and receives official approval it.

Tools of Mitigation

Four sets of tools that could be used to prevent or mitigate disasters include:

- Hazard management and vulnerability reduction.

- Economic diversification.
- Political intervention and commitment.
- Public awareness.

The first two apply exclusively to disasters caused by natural phenomena while the latter are used to mitigate any other hazards.

Importance of Mitigation

The significance of mitigation arises from the fact that any effort in preempting the occurrence of disaster, saves resources be it personnel, finance, physical assets, etc. The huge amount of financial resources allocated to mitigation activities reduces the demand for money towards emergency recovery, repair and reconstruction. Mitigation, since it aims at prevention and preparedness, ensures socio-economic continuity, sustainability and socio environmental health of the community.

The Yokohama message emanating from the International Decade for Natural Disaster Reduction in May, 1994, underlined the need for an emphatic shift in the strategy for disaster mitigation. It was inter alia stressed that disaster prevention, mitigation, preparedness and relief are four elements, which contribute to and gain, from the implementation or the sustainable development policies. These elements along with environmental protection and sustainable development, are closely interrelated. Therefore, nations should incorporate them in their development plans and ensure efficient follow up measures at the community, sub-regional, regional, national and international levels. The Yokohama Strategy also emphasised that disaster prevention, mitigation and preparedness are better than disaster response in achieving the goals and objectives of vulnerability reduction. Disaster response alone is not sufficient as it yields only temporary results at a very high cost. Prevention and mitigation contribute to lasting improvement in safety and are essential to integrated disaster management.

Strategies of Mitigation

The natural hazard management process can be divided into pre-event measures, actions during and immediately following an event, and post-disaster measures. In approximate chronological order, these are as follows:

- Pre-event Measures:
  - Mitigation of Natural Hazards:
    - Data Collection and Analysis,
    - Vulnerability Reduction.

    - Preparation for Natural Disasters:
        - Prediction,
        - Emergency preparedness (including monitoring, alert, evacuation),
        - Education and Training.
- Measures During and Immediately after Natural Disasters:
    - Rescue,
    - Relief.
- Post-disaster Measures:
    - Rehabilitation,
    - Reconstruction.

Of these, the mitigation mechanisms are most cost effective in reducing loss of life and property and most compatible with the development planning process. The data collection effort refers to data on the hazards themselves, vulnerability, and risk. The mitigation mechanisms are described briefly below.

Natural Hazard Assessments

Studies that assess hazards provide information on the probable location and severity of dangerous natural phenomena and the likelihood of their occurring within a specific time period in a given area. These studies rely heavily on available scientific information, including geologic, geomorphic, and soil maps; climate and hydrological data; and topographic maps, aerial photographs, and satellite imagery. Historical information, both written reports and oral accounts from long-term residents, also helps characterize potential hazardous events. Ideally, a natural hazard assessment promotes an awareness of the issue in a developing region, evaluates the threat of natural hazards, identifies the additional information needed for a definitive evaluation, and recommends appropriate means of obtaining it.

Vulnerability Assessments

Vulnerability studies estimate the degree of loss or damage that would result from the occurrence of a natural phenomenon of given severity. The elements analyzed include human populations; capital facilities and resources such as settlements, lifelines, production facilities, public assembly facilities, and cultural patrimony; and economic activities and the normal functioning of settlements. Vulnerability can be estimated for selected geographic areas, e.g., areas with the greatest development potential or already developed areas in hazardous zones. The techniques employed include lifeline (or critical facilities) mapping and sectoral vulnerability analyses for sectors such as energy, transport, agriculture, tourism, and housing. In Latin America and the Caribbean vulnerability to natural hazards is rarely considered in evaluating an investment even though vulnerability to other risks, such as fluctuating market prices and raw-material costs, is taken into account as standard practice.

Risk Assessments

Information from the analysis of an area's hazards and its vulnerability to them is integrated in an analysis of risk, which is an estimate of the probability of expected loss for a given hazardous event. Formal risk analyses are time-consuming and costly, but shortcut methods are available which give adequate results for project evaluation. Once risks are assessed, planners have the basis for incorporating mitigation measures into the design of investment projects and for comparing project versus no-project costs and benefits.

Vulnerability Reduction

Risk from natural hazards can be substantially reduced by the introduction of mitigation measures, both structural and non-structural.

Incorporating Mitigation Measures into an Integrated Development Planning Study

Integrated development planning is a multidisciplinary, multisectoral approach to planning. Issues in the relevant economic and social sectors are brought together and analyzed vis-a-vis the needs of the population and the problems and opportunities of the associated natural resource base. A key element of this process is the generation of investment projects, defined as an investment of capital to create assets capable of generating a stream of benefits over time. A project may be independent or part of a package of projects comprising an integrated development effort. The process of generating projects is called the project cycle. This process proceeds from the establishment of development policies and strategies, the identification of project ideas, and the preparation of project profiles through prefeasibility and feasibility analyses (and, for large projects, design studies) to final project approval, financing, implementation and operation.

While the process is more or less standardized, each agency develops its own version. The development planning process evolved by the OAS Department of Regional Development consists of four stages: Preliminary Mission, Phase I (development diagnosis), Phase II (project formulation and preparation of an action plan), and implementation. Because the process is cyclical, activities relating to more than one stage can take place at the same time. The main elements of the process are shown in figure, and a synthesis of the activities and products of each stage is shown in figure.

This presentation of the procedures of an integrated study features the incorporation of hazard management considerations at each stage.

Generally, planners depend on the science and engineering community to provide the required information for natural hazard assessments. If the information available is adequate, the planner may decide to make an assessment. If it is not adequate, the planner usually decides that the time and cost of generating more would be excessive, and the assessment is not made. While the information available on hurricanes and geologic hazards is often adequate for a preliminary evaluation, the information on desertification, flooding, and landslide hazards rarely is. The OAS has developed fast, low-cost methodologies that make these evaluations possible in the context of a development study. The differences in treating the various hazards in each stage of the process are highlighted in the following discussion.

Preliminary Mission: Designing the Study

The first step in the process of technical assistance for an integrated development planning study is to send a "preliminary mission" to consult with officials in the interested country. Experience has shown that this joint effort of OAS staff and local planners and decision-makers is frequently the most critical event in the entire study. They take action to:

Determine whether the study area is affected by one or more natural hazards. For example, the National Environmental Study of Uruguay conducted by the OAS with financial support from the Inter-American Development Bank determined in the preliminary mission that natural hazards were an important environmental problem, and consequently an assessment of all significant hazards, to be conducted by reviewing existing information, was programmed for Phase I.

- Identify the information available for judging the threat posed by those hazards in the study area: history of hazardous events; disaster and damage reports; assessments of hazards, vulnerability, risk; maps and reports on natural resources and hazards; topographic maps, aerial photographs, satellite imagery.
- Determine whether the available data are sufficient to evaluate the threat of hazards. If they are not, determine what additional data collection, hazard assessment, remote sensing, or specialized equipment will be needed for the next stage of the study. For example, in preliminary missions in Dominica, Saint Lucia, and St. Vincent and the Grenadines, landslides were determined to be a serious problem, and landslide assessments were included in the work plan for Phase I.
- Determine whether the studies required would serve more than one sector or project. If so, establish coordination.
- Establish coordination with the national institution responsible for disaster planning.

- Prepare an integrated work plan for Phase I that specifies the hazard work to be done, the expertise needed, and the time and cost requirements.

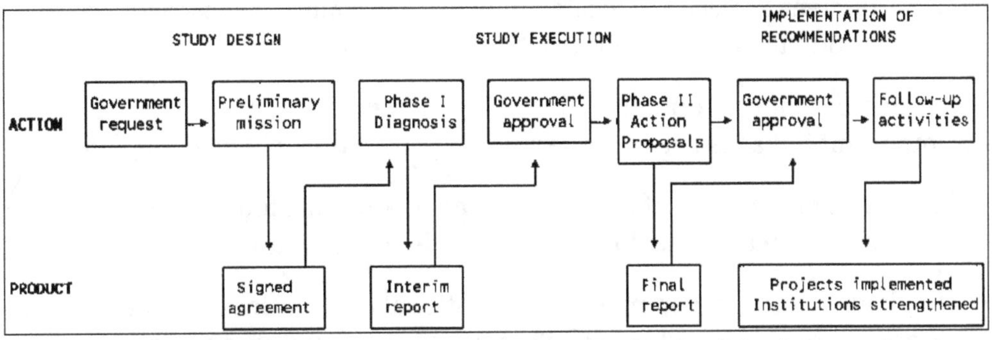

Key elements in the process of oas assistance for integrated regional development planning

Phase I: Development Diagnosis

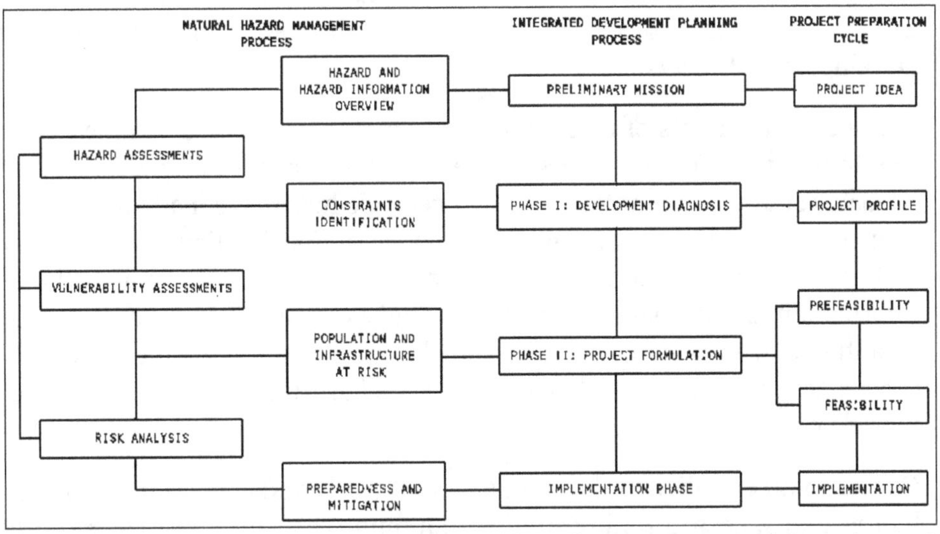

Integrated development planning process, natural hazard management, and the project cycle.

In Phase I, the team analyzes the study region and arrives at detailed estimates of development potentials and problems of the region and selected target areas. From this analysis a multisectoral development strategy and a set of project profiles are prepared for review by government decision-makers. Phase I also includes a detailed assessment of natural hazards and the elements at risk in highly vulnerable areas which facilitates the early introduction of non-structural mitigation measures. During this phase the team will:

- Prepare a base map.
- Determine the goods, services, and hazards of the region's ecosystems. Identify cause-and-effect relationships between natural events and between natural events and human activity. In the hilly Chixoy region of Guatemala, for example, it was found that inappropriate road construction methods were causing landslides and

that landslides, in turn, were the main problem of road maintenance. In Ecuador, the discovery that most of the infrastructure planned for the Manabí Water Development Project was located in one of the country's most active earthquake zones prompted a major reorientation of the project.

- Evaluate socioeconomic conditions and institutional capacity. Determine the important linkages between the study region and neighboring regions.
- Delineate target areas of high development potential, followed by more detailed natural resource and socioeconomic studies of these areas.
- In planning the development of multinational river basins or border areas where a natural disaster could precipitate an international dispute, make an overall hazard assessment as part of the resource evaluation. Examples of such studies include those for the development of the San Miguel-Putumayo River Basin, conducted in support of the Colombia-Ecuador Joint Commission of the Amazon Cooperation Project, and for the Dominican Republic and Haiti Frontier Development Projects.
- Conduct assessments of natural hazards determined to be a significant threat in the study region. For hurricanes and geologic hazards, the existing information will probably suffice; if the information on geologic hazards is inadequate, an outside agency should be asked to conduct an analysis. For flooding, landslides, and desertification, the planning team itself should be able to supplement the existing information and prepare analyses. The studies of the Honduran departments of Atlántida and Islas de la Bahía included flood hazard assessment as part of the coastal area development plan and landslide hazard assessments for some of the inland areas.
- Conduct vulnerability studies for specific hazards and economic sectors. Prepare lifeline maps, hazard zoning studies, and multiple hazard maps as required. The study of the vulnerability of the Ecuadorian agriculture sector to natural hazards and of ways to reduce the vulnerability of lifelines in St. Kitts and Nevis, for example, both generated project ideas which could be studied at the prefeasibility level in Phase II. The study of the Paraguayan Chaco included flood and desertification assessments and multiple-hazard zoning. The execution of these hazard-related activities did not distort the time or cost of the development diagnosis.
- Identify hazard-prone areas where intensive use should be avoided.
- Prepare a development strategy, including non-structural mitigation measures as appropriate.
- Identify project ideas and prepare project profiles that address the problems and opportunities and that are compatible with political, economic, and institutional constraints and with the resources and time frame of the study.

- Identify structural mitigation measures that should be incorporated into existing facilities and proposed projects.
- Prepare an integrated work plan for the next stage that includes hazard considerations.

Phase II: Project Formulation and Action Plan Preparation

At the end of Phase I a development strategy and a set of project profiles are submitted to the government. Phase II begins after the government decides which projects merit further study. The team now makes prefeasibility and feasibility analyses of the projects selected. Refined estimates are made of benefits (income stream, increases in production, generation of employment, etc.) and costs (construction, operation and maintenance, depletion of resources, pollution effects, etc.). Valuative criteria are applied, including net present value, internal rate of return, cost-benefit ratio, and repayment possibilities. Finally, the team assembles packages of investment projects for priority areas and prepares an action plan.

- Examine the human activities that could contribute to natural hazards (e.g., irrigation, plowing in the dry season, and animal husbandry could cause or exacerbate desertification) and the social and cultural factors that could influence project vulnerability during and after implementation.
- Determine the levels of technology, credit, knowledge, information, marketing, etc., that it is realistic to expect will be available to the users of the land, and ensure that the projects formulated are based on these levels.
- Prepare site-specific vulnerability and risk assessments and appropriate vulnerability reduction measures for all projects being formulated. For example, the multimillion-dollar program for the development of the metropolitan area of Tegucigalpa, Honduras, featured landslide mitigation components. Flood alert and control projects were central elements in the comprehensive Water Resource Management and Flood Disaster Reconstruction Project for Alagoas, Brazil.
- Mitigate the undesirable effects of the projects, avoid development in susceptible areas, recommend adjustments to existing land use and restrictions for future land use.
- Examine carefully the compatibility of all projects and proposals.
- Define the specific instruments of policy and management required for the implementation of the overall strategy and the individual projects; design appropriate monitoring programs.

Implementing the Study Recommendations

The fourth stage of the development planning process helps implement the proposals by preparing the institutional, financial, and technical mechanisms necessary for successful execution and operation. Efforts made to consider hazards in previous stages will be lost unless mitigation measures are closely adhered to during the projects' execution. Either the planning agency or the implementing agency should:

- Ensure that suitable hazard management mechanisms have been included in all investment projects; provide for monitoring of construction to ensure compliance with regulations, and for ongoing monitoring to ensure long-term compliance with project design.
- Ensure that national disaster management organizations have access to the information generated by the study. Point out hazardous situations for which the study did not propose vulnerability reduction measures.
- Arrange for the continuing collection of hazard data and the updating of information of planning and emergency preparedness agencies.
- Prepare legislation mandating zoning codes and restrictions, building and grading regulations, and any other legal mechanisms required.
- Include adequate financing for hazard mitigation measures.
- Involve the private sector in the vulnerability reduction program.
- For community-based vulnerability reduction programs, establish national training and hazard awareness programs for town and village residents, a feature of OAS technical assistance programs for Saint Lucia and Grenada.
- Generate broad-based political support through the media, training programs, and contacts with community organizations. Use products of the studies (photos, maps, charts, etc.) for mass communication. Use personnel who participated in the studies in public meetings to promote the concept of vulnerability reduction.
- Accelerate the implementation of projects that include hazard mitigation considerations; if budget cutbacks occur, reduce the number of projects rather than dropping the hazard mitigation components.

Advantages of Integrated Development Planning for Hazard Management

Even though integrated development planning and hazard management are usually treated in Latin America and the Caribbean as parallel processes that intermix little with each other, it is clear that they should be able to operate more effectively in coordination, since their goals are the same-the protection of investment and improved human well-being-and they deal with similar units of space. Some of the advantages of such coordination are the following:

- There is a greater possibility that vulnerability reduction measures will be implemented if they are a part of development package. The possibility increases if they are part of specific development projects rather than stand-alone disaster mitigation proposals. Furthermore, including vulnerability reduction components in a development project can improve the cost-benefit of the overall project if risk considerations are included in the evaluation. A dramatic example is the case study on vulnerability reduction for the energy sector in Costa Rica.
- Joint activities will result in a more efficient generation and use of data. For example, geographic information systems created for hazard management purposes can serve more general planning needs.
- The cost of vulnerability reduction is less when it is a feature of the original project formulation than when it is incorporated later as a modification of the project or an "add-on" in response to a "hazard impact analysis." It is even more costly when it is treated as a separate "hazard project," independent of the original development project, because of the duplication in personnel, information, and equipment.
- Exchanging information between planning and emergency preparedness agencies strengthens the work of the former and alerts the latter to elements whose vulnerability will not be reduced by the proposed development activities. In the Jamaica study of the vulnerability of the tourism sector to natural hazards, for example, solutions were proposed for most of the problems identified, but no economically viable solutions were found for others. The industry and the national emergency preparedness agency were so warned.
- With its comprehensive view of data needs and availability, the planning community can help set the research agenda of the science and engineering community. For example, when a planning team determines that a volcano with short-term periodicity located close to a population center is not being monitored, it can recommend a change in the priorities of the agency responsible.
- Incorporating vulnerability reduction into development projects builds in resiliency for the segment of the population least able to demand vulnerability reduction as an independent activity. A clear example of this situation was the landslide mitigation components of the metropolitan Tegucigalpa study: the principal beneficiaries were the thousands of the city's poor living in the most hazard-prone areas.

Guidelines for Planning in Re-building Process

In a post disaster situation, often the scale and urgency mean that despite good intentions, some key issues can get overlooked.

This resource pack has been put together to provide guidance on how to ensure that this doesn't happen.

A checklist for essential issues to be covered in an assessment [1]

- Household information of the communities
- Livelihood patterns
- Natural resource analysis
- Institutional structures
- Community/social support structures
- Overall development concerns and priorities
- Hazard analysis
- Risk analysis
- Vulnerability and capacity analysis
- Existing preparedness and emergency management plans and strengths
- Community coping mechanisms
- Gender issues and concerns

A checklist for assessment of a risk-area

- Livelihood patterns and their dynamics
- Resources and assets of the people in the community
- Availability of water
- Availability and supply of food
- Availability of fodder for livestock
- The formal institutional structures and accessibility
- Health and education services as well as their accessibility and availability to the people
- Quality and effectiveness of health services
- Shelter opportunities and constraints in the area
- Means of communication in the area
- Employment opportunities available in the community
- Available skills of the people, which can provide alternative employment
- Potential and constraints to promoting those skills
- Social networks which can provide safety-nets to the people during vulnerable times
- Management practices of Common Property Resources
- Governmental policy towards relief and development activities in the area
- The work of social organizations in the communities

Planning for reconstruction involves:

Physical reconstruction: (e.g. of buildings, infrastructure etc.)

As the urgent and immediate basic needs of affected communities are being met, the process of rehabilitation and reconstruction needs to begin. In order to ensure that rebuilding is done in a way that reflects the needs and capacities of affected communities the following checklists will be helpful:

This needs to be based on the hazard risk, vulnerability and capacity assessments. For example planning for housing will need to look at any hazards of the particular location, what kind of resources and capacities people have to be involved in the rebuilding, and what designs are appropriate like flood or drought resistant. Planning should also focus on needs of women, disabled and elderly.

Community Rehabilitation

This involves mobilising people and making them comfortable to live in their environment again, such as through setting up support groups, vigilant group for early warning and awareness building. This also means making sure that infrastructure ad service plans address needs e.g. buildings are safe, communities have access to information etc.

Livelihood Rebuilding

Plans should not only focus on the main affected sectors (fisheries, tourism, agriculture), but should also address the knock-on effects to other sectors. Livelihood rebuilding should recognise the needs and seek ways to support individuals and families with special needs such as where people have to take care of injured or children while also earning a living (e.g. investigate home-based work options for single headed households; make efforts to ensure inclusion of people with disabilities in training).

Rebuilding should also include innovative opportunities for people who didn't have options before, e.g. unemployed youth. The rebuilding process should in fact expand opportunities for poor/marginalized groups through innovative strategies to engage them in infrastructure redevelopment and other rebuilding activities. Rebuilding should also ensure that issues and concerns of communities prior to the disaster are addressed.

Using Hazard, Risk and Vulnerability Assessment and Capacity Assessment in Planning for Rebuilding

In the affected communities:

- Conduct a hazard risk assessment.
- Conduct a vulnerability assessment.
- Conduct a capacity assessment.

The following information is extracted from the Community based Disaster Management Course – Trainers Guide, Partnerships for Disaster Reduction in Southeast

Asia, Asian Disaster Preparedness Centre and European Community Humanitarian Office.

Nature and Behavior of a Hazard

- Force: Wind, water (rain, flood, overflow, run-off, flash flood tidal wave, storm surge, epidemic). Land (slides, deposits by river, lahar, mudflow) fire (forest fire, settlement fire), seismic (earthquake, tsunami, liquefaction), conflicts (civil war, insurgency, other actions leading to displacement and refugees), industrial/technological (pollution, radio-activity, explosions), other human-related (famine, drought, pests, etc.)
- Warning signs and signals: Scientific and indigenous indicators that hazard is likely to happen.
- Forewarning: Time between warning and impact.
- Speed of onset: Rapidity of arrival and impact. We can distinguish between hazards that occur without almost any warning (earthquake), and hazards that can be predicted three to four days in advance (typhoon) to very slow-onset hazards like drought and famine.
- Frequency: Does hazard occur seasonally, yearly, once every 10 years, once in a lifetime, etc.
- When: Does hazard occur at a particular time of the year (wet or dry season; in November to December).
- Duration: How long is hazard felt (earthquake and aftershocks; days/week/ months that area is flooded; length of period of military operations.

Guidelines for Conducting Vulnerability Assessment

Vulnerability Assessment is the process of estimating the vulnerability to potential hazards by:

- Identifying what elements are at risk per hazard type.
- Analysing the root causes of why these elements are at risk.

Vulnerabilities can be distinguished in three types (Refer to M2-04 'Disaster Crunch Model'):

- Physical/material.
- Social/organisational.
- Motivational/attitudinal.

Physical/Material Vulnerability

- Disaster-prone location of community, houses, farmlands, infrastructure, basic services, etc.
- Insecure sources of livelihood.
- Risky sources of livelihood.
- Lack of access and control over means of production (land, farm inputs, animals, capital, etc.
- Dependent on money-lenders, usurers, etc.
- Inadequate economic fall-back mechanisms.
- Occurrence of acute or chronic food shortage.
- Lack of adequate skills and educational background.
- Lack of basic services: education, health, safe drinking water, shelter, sanitation, roads, electricity, communication, etc.
- High mortality rates, malnutrition, occurrence of diseases, insufficient caring capacity.
- Overexploited natural resources.
- Exposed to violence (domestic, community conflicts, or war).

Social/organisational Vulnerability

- Weak family/kinship structures.
- Lack of leadership, initiative, organisational structures to solve problems or conflicts.
- Ineffective decision-making, people/groups are left out, etc.
- Unequal participation in community affairs.
- Rumours, divisions, conflicts: ethnic, class, religion, cast, ideology etc.
- Injustice practices, lack of practices to political processes.
- Absence of weak community organisations ((in) formal, governmental, indigenous).
- No or neglected relationship with government, administrative structures.
- Isolated from outside world.

Motivational/Attitudinal Vulnerability

- Negative attitude towards change.
- Passivity, fatalism, hopelessness, dependant.
- Lack of initiative no 'fighting spirit'.
- Lack of unity, cooperation, solidarity.
- Negative beliefs/solidarity.
- Unawareness about hazards and consequences.
- Dependence on external support/dole-out mentality.

Guidelines for Conducting Capacity Assessment

Capacity assessment is the process to determine what people do in times of crisis to reduce the damaging effects of the hazard, and to secure the sustainability of their livelihood by:

- Understanding people's previous experiences with hazards that enabled them to develop coping strategies.
- Analysing which resources are available and used by the community to reduce risk, who has access to these resources and who controls them.
- Assessing capacities of people at risk is a very important step in choolsing strategies for community disaster risk reduction and capability building. It is a step in the risk assessment process that most people forget. When we leave it out, we can make mistakes in program design and waste scarce external resources. Furthermore, if we ignore rather than support the strengths of the community, the output is that their coping mechanisms might be undermined and weakened, and eventually increase their vulnerability.

Exercise to help participants identify resources available for short-term risk reduction both at the individual and at community level (1 hour). In this activity participants use themselves and the communities where they work as reference point. They brainstorm resources available for reducing risk, categorise them and discuss them in terms of value, access and control.

- Outline the purpose and the process of the exercise.
- Ask participants to do brief buzz session with the person next to them, on the following:
    - What does 'capacity' mean? Think of M2-03:Definitions and concepts.
    - What does the term 'resources' mean to you?

- What does 'coping' mean?

In plenary, clarify the term and build on common understanding:

- Stress that likewise the vulnerability assessment, we distinguish three kind of capacities: physical/material; social/organisational; motivational/attitudinal.

Engaging the Community

Indigenous and time tested knowledge, informal networks and mechanisms for interaction, community support mechanisms, ways and means of linking community knowledge and capacities with the formal institutional structures, enhancing community capacities to cope need to be incorporated in plans:

Engaging the Community

- Identify and list the most successful strategies used by communities living in disaster-prone areas.
- In consultation with the community, build on indigenous measures to develop strategies that have already been successful. The following mechanisms are recommended to integrate community participation in the planning process:
  - Include both women and men from the community in the assessment, planning, and implementing of preventive measures.
  - Invite women and men who are leaders in the community to strategic planning meetings and discussions, to enable a better focus on ground realities, leading to more targeted strategies.
  - Identify and plan out the most useful forms of communication that can highlight and address the needs and concerns of women and men.
  - Conduct workshops with children: girls and boys; women and men to devise posters, street drama, songs, plays, etc. on disaster preparedness.
  - Invite schools, community-based organizations, and women's organizations to develop materials to inform the public about disaster preparedness and emergency management.
  - Bring together the different village-level organizations working in the area and share the focus / mandate of these organizations.
  - Make sure that community-level practitioners/ CBOs secure the involvement of both women and men in the community in local organizations.

Paying Attention to Women's and Gender Issues in Responding to the Tsunami Crisis

The recent tsunami disaster has resulted in many deaths, displacement of thousands and destruction of livelihoods, infrastructure and property. Given the scale and complexity of the situation there is a rush to attend to immediate needs to restore to normalcy, to initiate rehabilitation and reconstruction. Numerous agencies; government, UN, INGOs, local NGOs, and individuals are attending to the immediate relief measures, and soon a massive reconstruction and rehabilitation process will begin. Some will have previous experience and be knowledgeable and skilled in disaster situations. There are also many organisations/groups and volunteers new to crisis situations who are trying to help.

Despite good intentions, there is always the risk that important issues get bypassed. Experience shows that gender, in particular addressing women's issues in disaster situations, is a key area where there will be gaps unless given specific attention.

Although women and men have many common concerns, disasters do affect women and men differently: because of the different roles they occupy in society/community; the different and responsibilities given to them in life; and because of the differences in their capacities, needs, and vulnerabilities.

Ignorance of gender differences leads to insensitive and ineffective operations that largely bypass women's needs and their potential to assist in disaster relief and reconstruction activities.

This note highlights important messages about addressing women's concerns and gender issues in relief and reconstruction.

Making Initial Disaster Responses Gender Sensitive

Relief Distribution

Disaster relief that is gender sensitive requires:

- Close interaction with the affected communities during the relief planning process.
- Gender-disaggregated assessments for relief distribution.
- Employment of female relief workers.
    - E. g. distributing provisions through women.
- Relief workers who are aware and sensitive to gender issues and humanitarian ethics.
- Recognition of skills and capacities of women from affected communities and their involvement in relief planning, distribution of assistance and in other emergency management activities.
- Relief that reaches sub-categories such as widows, old women, female-headed households, single women, disabled etc.
- Attention to the cultural concerns of different communities and elimination of culture/religion/gender based discrimination in registration, compensation and relief distribution.
- Adherence to the minimum standards set for relief distribution (e.g. the SPHERE standards).

Basic Practical Needs

Women have specific needs; and measures should be taken to:

- Ensure privacy for women in common areas of camps.
    - E. g. provide women's "corners", separate toileting and bathing areas.
- Provide sanitary ware and rags for menstruation, and clothing such as undergarments.
- Attend to needs of pregnant and nursing mothers:
    - E. g. provide infant milk powder, feeding bottles, infant clothing, nappies and mosquito nets.

Security and Safety

It is a fact that in displaced situations, in temporary shelter and in camps, women and children are often subject to sexual harassment, abuse and violence. Specific measures need to be taken to secure women and children's safety:

- Take practical measures to protect them from abuse:
    - E. g. secure sleeping arrangements, adequate lighting and safe location of toilets.
- Take steps to ensure that the community is responsible for the safety of children.
- Where possible, assist and accompany women/children going in search of loved ones.

Health Concerns

Women keep families healthy after disasters. As caregivers to the young, old, sick, disabled, and injured, women tend to put their own needs last. Relief and reconstruction efforts need to pay attention to women's health and ensure specific health concerns and needs are being addressed:

- Measures are needed to tackle the increased risk and incidence of sexual and domestic violence associated with major disasters.
    - E. g. medical assistance should be available to women and children victims of physical or sexual abuse. Some women may need the morning after pill.
- Reproductive and family planning health services should be included in general health work.

- E. g. provision made for antenatal and postnatal care; pregnant and lactating women who may need nutritional supplements.

- The different physical and mental health needs of women and men need to be recognised and addressed.
  - E. g. people with disabilities, elderly people and family care givers.

Trauma Counseling

Members of relief teams need to be aware and sensitive to the issues of trauma:

- Gender differences in psychological impacts of disasters should recognise that women's anxiety also stems from fear and risk to their family/children.
- Training for mental health providers should address problems of highly vulnerable groups such as women headed households, grandmothers caring for orphans, battered women, women with disabling injuries, newly widowed women and men, women at risk of suicide.

Gender Sensitive Planning for Rehabilitation/reconstruction

In many communities, women take an active part in community disaster initiatives. Yet in larger, more formal planning, women are scarcely represented and markedly absent from decision-making. Not being sensitive to gender issues in development planning and disaster mitigation means that interventions are often only targeted at men. Sensitivity to gender is vital in order to empower a community to successfully move on and move up from the abyss of disaster.

Rehabilitation/reconstruction should promote post-disaster development that reduces risk of communities to disaster and empowers local communities. This means tackling the reasons why certain sections of society and community are more vulnerable to disasters. Rebuilding should happen in ways that address the root causes of vulnerability, including gender inequalities.

Women's local knowledge and expertise are essential assets for communities and households struggling in to rebuild. To capture these capacities, disaster responders must work closely with women. In planning and implementation of rehabilitation/reconstruction, practical steps should be taken to:

- Ensure the needs, skills and capacities of affected communities are incorporated in planning and implementing rehabilitation work.
  - E. g. include women in housing design as well as construction; recognise and incorporate women's traditional knowledge and experience in managing natural resources.

- Establish on-going consultation with women in affected areas, women's bureaus, and women's advocacy groups.
- Evaluate and take measures to ensure women can participate in reconstruction and benefit from economic recovery packages.
    - E. g. Ensure that women have the mobility to participate in reconstruction and rehabilitation activities. Ensure meetings and events are held at times and places where women can participate; Ensure family caregivers have access to access to support.
- Strengthen informal social networks and link them to disaster-responding agencies and offices.
- Fund women's groups to monitor disaster recovery projects.
- Identify and respond to women's needs for legal services in the areas of housing, employment, and family relations.
    - E. g. Deed newly constructed houses in both the names of husband and wife, and land rights for women.
- Give priority to social services, children's support systems and women's centres.
- Target highly vulnerable women such as single mothers, widows, below-poverty, unemployed women and socially marginalized women in reconstruction of damaged and new houses.
- Monitor relief and rehabilitation for possible gender bias and inequities that may develop over time.
    - E. g. avoid unintentional overburdening of women with multiple responsibilities at home, work, and in the community.
- Monitor as far as possible the degree to which relief and recovery assets are equitably distributed.

Re-building Livelihoods

Reconstruction must fully engage women and ensure that women benefit from economic recovery and income support programs. Women's limited income generation and employment opportunities should be expanded in the process of developing local economies. In re-building livelihoods, practical steps should be taken to:

- Ensure rehabilitation and reconstruction target economically active women of all ages and social groups.

- Incorporate gender analysis into all empirical assessments.
  - E. g. collect or generate gender-specific data; conduct a thorough analysis of damaged economic sectors (e.g. fishery, tourism, agriculture) that identifies roles of women and identify areas for their participation.
- Support income-generation projects that build non-traditional skills among women.
  - E. g. provide women with fair access to construction-related and other non-traditional employment; include employment-relevant job training; seek out women with technical qualifications for training on specific projects such as overseeing housing construction.
- Incorporate women's income generating options in livelihood rebuilding plans.
  - E. g. make provision for self-employed/home-based women workers in plans.
- Ensure access to grants and loans to re- build lost livelihoods to replace damaged or destroyed tools, workspace, equipment, supplies, credit, capital, markets and other economic resources.

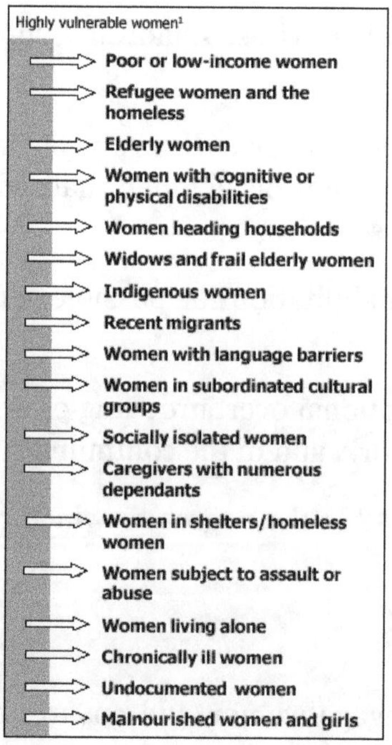

- Include measures to support women's multiple responsibilities as economic providers and family workers.
  - E. g. work with employers to develop or strengthen "family friendly" policies for those needing time to apply for assistance, cope with trauma and help injured

family members'; provide assistance to family care givers to support them economically and ensure continued care to the injured, children, and disabled.

- Develop and commit to gender accountability and monitoring measures.
    - E. g. monitor the percentage of women and men in construction, trade, other employment; the numbers of disabled women trained; the proportion of economic recovery grants and loans funds received by women; the working conditions in private and public relief work projects etc.; monitor and assess long term impacts on women and girls of disrupted markets, forced sale of assets, involuntary migration, increasing proportion of female- headed households etc.

Disability Sensitive Planning for Relief/Rehabilitation/ Reconstruction

The tsunami disaster is unprecedented, the most devastating natural disaster Sri Lanka has ever faced.

Apart from the loss of lives and property there will be more long term social catastrophes that will add to the tragedy such as the increased number of persons with disabilities (PWDs) and other vulnerable groups of such as single parented families, infants, children, and young adults left hopeless in these areas.

It is reported that there are more than 16,000 survivors with injuries. Though the number is not certain many of these people will sustain visible physical disabilities. Moreover estimated 15,000 pregnant mothers affected by the disaster are, now living in refugee camps. The trauma they under go, lack of proper nutrition and inadequate medical facilities will have a greater possibility for them to end up with a baby with a disability. Other concern would be the illnesses people suffer and diseases spreading in the affected areas which could lead to disability conditions. All these would add up to make the number of PWDs a phenomenal figure if these issues are not addressed in the in the immediate relief efforts as well as the long-term prevention plans.

The survivors sustaining disabilities would also be in a depressed mental condition due to their situation and the feeling of they have become a burden to the remaining members of their families. They will require support to overcome the distress of the dramatic change in their lives in addition to the medical services and assistive devices to improve their mobility. In the wake of more urgent and basic needs there might be the possibility of PWDs being neglected. On a longer term they will also need to acquire skills to engage in new livelihood activities in order to be a productive part of the society.

The situation would be the same among single parented families, children and young adults who have lost their breadwinners. Therefore the key issues of the more vulnerable groups in the long run, such as the persons with disabilities, women, children and single parented families should also be taken in to account in the immediate relief plans and in long term rebuilding and strengthening plans in order for them to relive their lives with out being neglected in the overall development process.

The marginalized communities' project of ITDG-South Asia, aims to propose strategies to address the key issues of the more vulnerable groups of the disaster affected communities.

Guidelines for Paying Attention to Persons with Disabilities in Disaster Relief

Relief Distribution

Identification of persons with disabilities (PWDs) in the relief planning and distribution process Collect information on:

- Number of people.
- Type of disability.
- Specific requirements of PWDs - Medical services, assistive devices, Nutrition.

Relief workers made aware of the disability conditions and paying attention to avoid possible marginalization of these groups in a disaster situation.

Getting the support of the community (women, men and children affected) to assist the PWDs. Educate the community in order for them to be sensitive to issues of disability.

Attention to see that relief reaches the PWDs to avoid discrimination.

Special attention focusing different disability groups to provide services assistance and care.

- Children with disabilities.
- Women with disabilities.

Efforts to communicate directly with PWDs as much as possible to identify their basic and specific need.

E.g. Come up with simple methods for communication. Try writing down or get the support of a family member to communicate a deaf person.

Give them time to express their needs.

Fulfilling basic Needs

Arranging separate areas of the camp if the PWDs are not comfortable being in the crowds or in common areas. Let them interact with the crowds, gradually.

Provide assistance for PWDs without life skills eg. Eating, using toilets and cleaning themselves. Make the common facilities and area accessible for PWDs as much as possible, with the available resources. If not possible arrange separate private areas for PWDs fulfil their basic requirements. Relief registration and distribution areas, Toilets, Bathing areas. Provide required assistive devices for PWDs in order them to be independent in their work.

Security and Safety of PWDs

PWDs have a grater chance for harassment, abuse or being neglected in camps and temporary shelters. In order avoid this sort of a situation:

- Educate the community to understand the disability condition of the PWDs in the camp and promote them to assist PWDs instead of teasing, sympathizing or neglecting.
- Form care-taking groups to assist PWDs and ensure their security including family members, relatives and volunteers.
- PWDs may also get frustrated or violent in an unfamiliar situation in the camps which may cause difficulties for the other residents. Try to make arrangements to avoid such situations.
- Finding people that they trust as caretakers.
- Providing facilities to comfort PWDs in the camps (calm surroundings, adequate light) Providing medical precautions if possible.
- Measures to protect incidences of sexual harassment of PWDs.

Health needs of PWDs

Immediate medical attention to PWDs in the camps and provide with drugs and therapy required.

Prevention of Future Disability

Poor nutrition, sanitation and medical attention could result many of the vulnerable groups to ended up with disability conditions if not the situations are addressed properly.

Attention to vulnerable groups including infants, children, pregnant/lactating mothers, sick, injured Providing nutritional supplements.

Medical Attention

Sanitation and hygiene measure to prevent diseases spreading.

Rehabilitation and Reconstruction Work Focusing PWDs

Ensure the issues of disability are incorporated in the rehabilitation and rebuilding work.

Getting the inputs of the experts in disability sector in the planning process. Representation of the disability sector organizations in field implementation and monitoring. Take measures to ensure that PWDs are benefited from the rehabilitation and reconstruction programs. Increase the awareness on the compensation programs/ economic recovery. Packages among PWDs, their family members, care takers, and organizations engaged in the rehabilitation work. Increase the accessibility of PWDs to these supports (Time, place and mode of communication to ensure the participation of PWDs).

Improving the mobility and accessibility of PWDs through appropriate designs Providing options of disability friendly houses, toilets.

Common compounds, public places, play areas etc. reconstructed with the consideration of accesibilty for PWDs. Facilitate PWDs to obtain legal services required. Obtaining lost legal documents. Priority in housing and rights for the lands (to secure future of some PWDs and as an asset to economically active PWDs).

Disability specific rehabilitation programs to increase the mobility and level of independency of PWDs, in order to make them economically productive.

Medical treatments and therapy required (Physiotherapy, Speech therapy), appropriate assistive devices, training on mobility, counselling.

(Immediate attention of rehabilitation for people acquired disability due to the disaster would result a higher improvement).

Strengthen the organizations and local networks engage in the rehabilitation and reconstruction work to establish on-going consultation with the PWDs.

Training programs to sensitize issues of disability in their line of work.

Develop links with disability sector organizations for expertise and support required.

Provide a platform for PWDs to voice their needs in through these networks (formation of community groups including PWDs to voice their needs).

Improving Livelihoods of PWDs

Identification of different income generation options in the area where PWDs could involve fully or for a part of the process.

Training programs for PWDs to develop their skills for employment (Self employment/ home based production/ other employment).

Promote PWDs to move to viable lively hood options from the traditional self employment work Provide technological support.

Adaptation of equipments and production process to be applicable to different disability conditions.

Improve the access to loans, grants and other technological facilities by advocating respective organizations.

Community Early Warning Systems in Place

Hazards, Vulnerability and Disasters

Natural hazards, such as storms, droughts, volcanic eruptions, or earthquakes, need not spell disaster. A disaster occurs only if a community or population is exposed to the natural hazard and cannot cope with its effects. Torrential rain in the middle of an ocean will not cause a disaster, but the same heavy rainfall on a vulnerable population – say a shanty town on the side of a hillside stripped of trees - may result in landslides and a huge loss of life. A minor drought may cause a famine if a region's agricultural production is highly stressed by civil war. A community that lacks an early warning system may sleep while volcanic ash clouds bear down upon them.

Vulnerability is the potent additive that mixes with natural hazards to cause disasters.

More than just a Prediction

A complete and effective early warning system comprises a chain of four elements, spanning a knowledge of the risks faced through to preparedness to act on early warning. Failure in any one part can mean failure of the whole system.

The Four Elements of Effective Early Warning Systems

Prior knowledge of the risks faced by communities	Technical monitoring and warning service for these risks	Dissemination of understandable warnings to those at risk	Knowledge and preparedness to act
Risks arise from both the hazards and the vulnerabilities that are present – what are the patterns and trends in these factors?	Is there a sound scientific basis for predicting the risks faced? Are the right things being monitored? Can accurate warnings be generated in timely fashion?	Do the warnings get to those at risk? Do people understand them? Do they contain useful information that enable proper responses?	Do communities understand their risks? Do they respect the warning service? Do they know how to react?

Pulling it all Together

Good early warning systems have strong linkages between the four elements. The major players concerned with the different elements meet regularly to ensure they understand all of the other components and what other parties need from them. Risk scenarios are constructed and reviewed. Specific responsibilities throughout the chain are agreed and implemented. Past events are studied and improvements are made to the early warning system. Manuals and procedures are agreed and published. Communities are consulted and information is disseminated. Operational procedures such as evacuations are practiced and tested.

Behind all of these activities lies a solid base of political support, laws and regulations, institutional responsibility, and trained people. Early warning systems are established and supported as a matter of policy. Preparedness to respond is engrained in society.

Understanding Concepts on Hazards, Disasters and Vulnerability

Seeing Disasters Differently

Disasters are usually seen as sudden events that result in death and destruction; and which require immediate, emergency relief. Natural hazards are usually cited as the cause of disasters and are sometimes viewed as 'acts of the gods' and divine punishment. To date, disasters have largely been the concern of environmental and geological scientists who study natural hazards, and the humanitarian institutions that have to deal with the human costs.

The term 'natural hazard' is often confused with the term 'natural disaster'. In order to understand disasters better, the relationship between these two key concepts and other related ideas must to be explored.

A hazard can be defined as a phenomenon that has the potential to cause injury to life, livelihoods and habitats. High winds, floods, landslides, droughts and earthquakes are all natural hazards. If unmanaged, that is, if natural hazards and their potential consequences are not prepared for, they become disasters that take lives and damage livelihoods.

Vulnerability is a set of conditions that affect the ability of countries, communities and individuals to prevent, mitigate, prepare for and respond to hazards. High levels of vulnerability increase the likelihood that natural hazards will be unmanaged and result in disaster.

Although most natural hazards have the potential to be very destructive, they do not always turn into disasters. It is the combination of a natural hazard and vulnerability that leads to a disaster.

"A disaster occurs when a significant number of vulnerable people experience a hazard and suffer severe damage and disruption of their livelihood system in such a way that recovery is unlikely without external aid."

As disasters continue to affect more and more people, it becomes increasingly clear that the social, economic and political forces of development processes contribute to disasters. Aside from the sudden hazards that arise from nature, people may also be made vulnerable due to the changes wrought by the unequal distribution of resources, social and political processes that perpetuate inequality, global macro-economic forces, and uncoordinated, ill-planned development. Consequently, there is growing recognition of a need to re-examine how we view and act towards disasters.

Understanding Vulnerability to Disaster

Some groups are more vulnerable than others. Vulnerability is not just poverty, but the poor tend to be the most vulnerable due to their lack of choices. The influences of both poverty and development processes on people's vulnerability to disasters are now well established. Class, caste, ethnicity, gender, disability and age are other factors affecting people's vulnerability.

Because vulnerability plays such an important part in why natural hazards become human disasters, it is worth spending some time to examine the characteristics of vulnerability.

Conditions of vulnerability are a combination of factors that include poor living conditions, lack of power, exposure to risk, and the lack of capacity to cope with shocks and adverse situations. It can be helpful to think of vulnerabilities in the following way:

- Physical vulnerabilities are the hazard prone locations of settlement, insecure and risky sources of livelihood, lack of access to basic production resources (such as land, farm inputs, and capital), lack of knowledge and information, and lack of access to basic services.
- Social vulnerabilities are reflected in the lack of institutional support structures and leadership, weak family and kinship relations, divisions and conflicts within communities, and the absence of decision-making powers.
- Attitudinal vulnerabilities are seen in dependency, resistance towards change, and other negative beliefs.

Poverty does not equal vulnerability, but being poor makes people more vulnerable to disasters because poor people lack the resources (physical, social, and knowledge based), to prepare for and respond to such threats and shocks as natural hazards.

Poor people often get locked in a cycle of vulnerability. Because they are poor, they become vulnerable. Because they are vulnerable, they are at great risk in the face of a natural hazard – leading to disaster. Because they suffer greater losses from a disaster, they become even poorer, more vulnerable, and are at an even greater risk of another disaster.

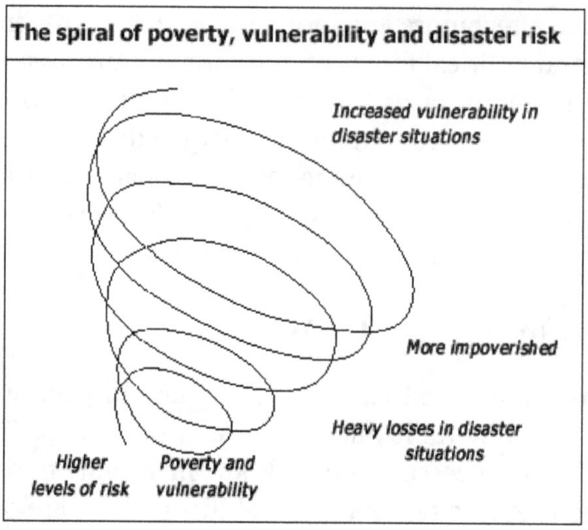

- Regardless of location, there is a direct link between poverty and powerlessness, and the severity of the impact of a disaster.
- Socially and economically vulnerable groups in any society are worst affected by natural hazards.
- There is a cycle of vulnerability to disasters – vulnerable groups are made even more vulnerable by disasters. This reduces their ability to pick up the pieces after a disaster and makes them further vulnerable in the face of another disaster.

References

- Disaster-recovery-control-measures- 40297: realtytimes.com, Retrieved 19 August, 2020
- Mitigation, emergency-management-emergencies-disasters: newbedford-ma.gov, Retrieved 23 April, 2020
- Importance-and-aims-of-disaster-mitigation-1742: preservearticles.com, Retrieved 31 March, 2020
- Mitigation-actions: mitigationguide.org, Retrieved 15 May, 2020
- What-hazard-mitigation: emergency.lacity.org, Retrieved 18 January, 2020
- Mitigation-in-disaster-management-meaning-examples-and-tools: gktoday.in, Retrieved 26 June, 2020

5
Warning Systems

Fire alarm system, emergency population warning system, emergency alert system, tsunami warning system, earthquake warning system, gale warning system, etc. are some of the warning systems that are used in emergency management. This chapter closely examines the different warning systems used in emergency management for a thorough understanding of the subject.

An Early Warning System (EWS) can be defined as a set of capacities needed to generate and disseminate timely and meaningful warning information of the possible extreme events or disasters (e.g. floods, drought, fire, earthquake and tsunamis) that threatens people‘s lives. The purpose of this information is to enable individuals, communities and organizations threatened to prepare and act appropriately and in sufficient time to reduce the possibility of harm, loss or risk.

Elements of Early Warning

Early warning is the integration of four main elements:

- Risk Knowledge: Risk assessment provides essential information to set priorities for mitigation and prevention strategies and designing early warning systems.
- Monitoring and Predicting: Systems with monitoring and predicting capabilities provide timely estimates of the potential risk faced by communities, economies and the environment.
- Disseminating Information: Communication systems are needed for delivering warning messages to the potentially affected locations to alert local and regional governmental agencies. The messages need to be reliable, synthetic and simple to be understood by authorities and public.
- Response: Coordination, good governance and appropriate action plans are a key point in effective early warning. Likewise, public awareness and education are critical aspects of disaster mitigation.

The purpose of early warning systems is to detect, forecast, and when necessary, issue alerts related to impending hazard events. In order to fulfill a risk reduction function, however, early warning needs to be supported by information about the actual and potential risks that a hazard poses, as well as the measures people can take to prepare for and mitigate its adverse impacts. Early warning information needs to be communicated in people friendly manner in such a way that facilitates decision-making and timely action of response organizations and vulnerable groups. Early warning information comes from different meteorological offices (for weather related disasters- flood, cyclone etc.); Ministries of Health (for example, disease outbreaks) and Agriculture (for example, crop forecasts); local and indigenous sources; media sources and increasingly from Internet early warning services.

Need of Early Warning System

Early Warning for disaster reduction is a legitimate matter of public policy at the highest national levels for two main reasons:

- The first one, clearly, is public safety, and the protection of human lives.
- The second is the protection of the nation's resource base and productive assets (infrastructure and private property or investments) to ensure long term development and economic growth. Conversely, by reducing the impact of disasters, a government avoids the financial and political- burden of massive rehabilitation costs.

Investing in early warning and other measures of disaster reduction is neither simple nor inexpensive, but the benefits of doing so, and the costs of failing to, are considerable. For instance:

- In terms of reducing economic losses, early warning and disaster preparedness pay for themselves' many times over the life of the warning system.
- The reduction of environmental losses can, if properly managed and publicized, have both long-term benefits to the economy, and short-term benefits for the administration in-charge.
- A country can strengthen its stature and influence in international relations by a good handling of externalities', or indirect effects, on neighboring nations.

From a public policy viewpoint, early warning, disaster preparedness and prevention must be part of a single, well integrated process.

Communication of Early Warning Information

An effective early warning system needs an effective communication system. Early warning communication systems are made of two main components:

- Communication infrastructure hardware that must be reliable and robust, especially during the natural disasters.

- Appropriate and effective interactions among the main actors of the early warning process such as the scientific community, stakeholders, decision makers, the public, and the media.

Many communication tools are currently available for warning dissemination such as Short Message Service (SMS) (cellular phone text messaging), email, radio, TV, and web service. Information and communication technology (ICT) is a key element in early warning. ICT plays an important role in disaster communication and dissemination of information to organizations in charge of responding to warnings and to the public during and after a disaster. Redundancy of communication systems is essential for disaster management, while emergency power supplies and back-up systems are critical in order to avoid the collapse of communication systems after disasters occur.

Community based Early Warning System

Early warning systems have limitations in terms of saving lives if they are not combined with "people-centered" networks. To be effective, early warning systems must be understandable, trusted by and relevant to the communities that they serve. Warnings will have little value unless they reach the people most at risk, who need to be trained to respond appropriately to an approaching hazard.

Community-Based Early Warning Systems (CBEWS) are anchored in the communities and managed by the communities. It is based on a "people-centered" approach that empowers individuals and communities threatened by hazards to act in sufficient time and in an appropriate manner in a bid to reduce the possibility of personal injury, loss of life, damage to property, environment and loss of livelihood. It provides communities, practitioners and organizations involved in disaster risk management with advance information of risks that can be readily translated into prevention, preparedness and response actions. CBEWS helps to reduce economic losses by allowing people to better protect their assets and livelihood. Essential features of community-based early warning systems are:

- All community members especially the vulnerable groups should be involved at all stages of the CBEWS from designing to operating the systems, receiving the warning messages and responding to the warning.
- Measures taken should be based on the needs of everyone in the community including the most vulnerable segments of the community.
- The community members will own the process and system.
- CBEWS measures will enhance the capacity of the community members to deal with their situation.
- Meaningful participation in the decision-making process of EWS.

Early Warning Systems and Policy

For early warning systems to be effective, it is essential that they be integrated into policies for disaster mitigation. Good governance priorities include protecting the public from disasters through the implementation of disaster risk reduction policies. It is clear that natural phenomena cannot be prevented, but their human, socio-economic and environmental impacts can and should be minimized through appropriate measures, including risk and vulnerability reduction strategies, early warning, and appropriate action plans. Most often, these problems are given attention during or immediately after a disaster. Disaster risk reduction measures require long term plans and early warning should be seen as a strategy to effectively reduce the growing vulnerability of communities and assets.

The information provided by early warning systems enables authorities and institutions at various levels to immediately and effectively respond to a disaster. It is crucial that local government, local institutions, and communities be involved in the entire policy making process, so they are fully aware and prepared to respond with short and long-term action plans.

Key elements for successful implementation of early warning:

- Understand the most likely threats, likelihood of disasters and their potential consequences: Although natural disasters are not precisely predictable, they are most often generally foreseeable. In other words, there are many areas where the occurrence of floods is likely; one does not necessarily know exactly when, but one knows they will occur sooner or later. Many natural hazards can be foreseen, or anticipated, from past experience, the analysis of current patterns of land use, or population distribution.
- Establish proper priorities: To allocate scarce resources most wisely, decision makers must rely on the type of analysis above, and make the disaster management choices which have the highest _value‘, in terms of losses avoided. One common approach is to use the expected value criteria; that is, the likelihood of an event multiplied by the potential cost of this event if it occurred. Simply stated, it is a matter of giving priority to the worst-most likely‘ over the most benign-least likely‘ events.
- Developing institutional networks with clear responsibilities: Understanding the nature of natural hazards and related vulnerabilities, for early warning purposes, requires a combination of actors from several areas, such as science and research (including social sciences and cultural aspects), land use planning, environment, finance, development, education, health, energy, communications, transportation, labor and social security as well as national defence. On the other hand, a prompt and effective response to a disaster, based on early warning, implies that concerted action –managed by a higher authority–be taken

by specific types of institutions: civil defence or public safety personnel, power and other utility agencies or companies, public health authorities, etc. at levels ranging from the cabinet minister's to the community leaders.

- Establish or strengthen the legislative/legal framework and mechanisms: Just as for any other aspect of public policy, early warning systems, as well as other disaster reduction applications need to be motivated and based within governmental responsibilities, especially since response to disasters may require exceptional executive powers for a specific period of time but its success cannot be accomplished without the benefits of widespread decision making and the participation of many others.
- Developing effective communication strategies: The context of early warning system communications has two aspects; the hardware aspect relates to the maintenance of lifelines, i.e. the necessity to build or strengthen robust hazard-resistant communication systems; the software aspect relates to the maintenance of relationships, i.e. the need to establish and maintain effective links and working relationships among the actors involved in the early warning communication chain.
- Securing resources: A substantial amount of resources is needed to ensure monitoring, adequate early warning, concerted disaster reduction, and a return to normal life. To a great extent, the capacity to secure resources to do this versus undertaking a competing public program—depends on the quality and credibility of the overall system: understanding threats, clear priority setting and institutional networks, and appropriate legislative dialogue.

Fire Alarm System

A fire alarm system has a number of devices working together to detect and warn people through visual and audio appliances when smoke, fire, carbon monoxide or other emergencies are present. These alarms may be activated automatically from smoke detectors, and heat detectors or may also be activated via manual fire alarm activation devices such as manual call points or pull stations. Alarms can be either motorized bells or wall mountable sounders or horns. They can also be (speaker strobes) which sound an alarm, followed by a voice evacuation message which warns people inside the building not to use the elevators. Fire alarm sounders can be set to certain frequencies and different tones including low, medium and high, depending on the country and manufacturer of the device. Most fire alarm systems in Europe sound like a siren with alternating frequencies. Fire alarm electronic devices are known as horns in the United States and Canada, and can be either continuous or set to different codes. Fire alarm warning devices can also be set to different volume levels.

A fire alarm notification appliance that is used in the United States and Canada.

Design

After the fire protection goals are established – usually by referencing the minimum levels of protection mandated by the appropriate model building code, insurance agencies, and other authorities – the fire alarm designer undertakes to detail specific components, arrangements, and interfaces necessary to accomplish these goals. Equipment specifically manufactured for these purposes is selected and standardized installation methods are anticipated during the design.

- ISO 7240-14 is the international standard for Design, installation, commissioning and service of fire detection and fire alarm system in and around building. This standard was published in August 2013; Status, Published; Edition 1; Technical Committee ISO/TC 21/SC 3 Fire detection and Alarm system.

- NFPA 72, *The National Fire Alarm Code* is an established and widely used installation standard from United States. In Canada, the ULC is the standard for the fire system.

Last version 2019; Status, Published. This code is part of family standard NFPA.

- TS 54 -14 is a Technical Specification (CEN/TS) for Fire detection and fire alarm system - Part 14: Guidelines for planning, design, installation, commissioning, use and maintenance. This document has been prepared by Technical Committee CEN/TC72. This document is part of the EN 54 series of standards.This standard was published in October 2018; Status, Published.

There are national codes in each European country for planning, design, installation, commissioning, use and maintenance of fire detection system with additional requiments that are mentioned on TS 54 -14.

- Germany, Vds 2095.

- Italy, UNI 9795.
- France NF S61-936.
- Spain UNE 23007-14.
- United Kingdom BS 5839 Part 1.

Parts

A fire alarm control panel.

- Fire alarm control panel (FACP) AKA fire alarm control unit (FACU): This component, the hub of the system, monitors inputs and system integrity, controls outputs and relays information.
- Primary power supply: Commonly the non-switched 120 or 240-volt alternating current source supplied from a commercial power utility. In non-residential applications, a branch circuit is dedicated to the fire alarm system and its constituents. "Dedicated branch circuits" should not be confused with "Individual branch circuits" which supply energy to a single appliance.
- Secondary (backup) power supplies: This component, commonly consisting of sealed lead-acid storage batteries or other emergency sources including generators, is used to supply energy in the event of a primary power failure. The batteries can be either inside the bottom of the panel or inside a separate battery box installed near the panel.
- Initiating devices: These components act as inputs to the fire alarm control unit and are either manually or automatically activated. Examples would be devices such as pull stations, heat detectors, duct detectors, and smoke detectors. Heat and smoke detectors have different categories of both kinds. Some categories are beam, photoelectric, ionization, aspiration, and duct.

- Fire alarm notification appliance: This component uses energy supplied from the fire alarm system or other stored energy source, to inform the proximate persons of the need to take action, usually to evacuate. This is done by means of a pulsing incandescent light, flashing strobe light, electromechanical horn, siren, electronic horn, chime, bell, speaker, or a combination of these devices. Strobes are either made of a xenon tube (most common) or recently LEDs.
- Building safety interfaces: This interface allows the fire alarm system to control aspects of the built environment, prepare the building for fire, and control the spread of smoke fumes and fire by influencing air movement, lighting, process control, and human transport and exit.

Initiating Devices

- Manually actuated devices; also known as fire alarm boxes, manual pull stations, or simply pull stations, break glass stations, and (in Europe) call points. Devices for manual fire alarm activation are installed to be readily located (near the exits), identified, and operated. They are usually actuated by means of physical interaction, such as pulling a lever or breaking glass.
- Automatically actuated devices can take many forms intended to respond to any number of detectable physical changes associated with fire: convected thermal energy; heat detector, products of combustion; smoke detector, radiant energy; flame detector, combustion gases; fire gas detector, and release of extinguishing agents; water-flow detector. The newest innovations can use cameras and computer algorithms to analyze the visible effects of fire and movement in applications inappropriate for or hostile to other detection methods.

Notification Appliances

A speaker and a remote light.

- Notification Appliances utilize audible, visible, tactile, textual or even olfactory stimuli (odorizer) to alert the occupants of the need to evacuate or take action in the event of a fire or other emergency. Evacuation signals may consist of simple appliances that transmit uncoded information, coded appliances that transmit a predetermined pattern, and or appliances that transmit audible and visible textual information such as live or pre-recorded instructions, and illuminated message displays.

- In the United States, fire alarm evacuation signals generally consist of a standardized audible tone, with visual notification in all public and common use areas. Emergency signals are intended to be distinct and understandable to avoid confusion with other signals.

As per NFPA 72, 18.4.2 (2010 Edition)Temporal Code 3 is the standard audible notification in a modern system. It consists of a repeated 3-pulse cycle (.5s on .5s off .5s on .5s off .5s on 1.5s off). Voice Evacuation is the second most common audible in a modern system. Legacy systems, typically found in older schools and building have used continuous tones alongside other audible schema.

- In the United Kingdom, fire alarm evacuation signals generally consist of a two tone siren with visual notification in all public and common use areas. Some fire alarm devices have an alert signal which is generally used for schools for lesson changes, start of morning break, end of morning break, start of lunch break, end of lunch break and when the school day is over.

- Audible textual appliances, which are employed as part of a fire alarm system that includes Emergency Voice Alarm Communications (EVAC) capabilities. High-reliability speakers are used to notify the occupants of the need for action in connection with a fire or other emergency. These speakers are employed in large facilities where general undirected evacuation is considered impracticable or undesirable. The signals from the speakers are used to direct the occupant's response. The system may be controlled from one or more locations within the building known as Fire Wardens Stations, or from a single location designated as the building Fire Command Center. Speakers are automatically actuated by the fire alarm system in a fire event, and following a pre-alert tone, selected groups of speakers may transmit one or more prerecorded messages directing the occupants to safety. These messages may be repeated in one or more languages. Trained personnel activating and speaking into a dedicated microphone can suppress the replay of automated messages in order to initiate or relay real-time voice instructions.

Emergency Voice Alarm Communication Systems

- Some fire alarm systems utilize emergency voice alarm communication systems (EVAC) to provide pre-recorded and manual voice messages. Voice alarm systems are typically used in high-rise buildings, arenas and other large "defend-in-place" occupancies such as hospitals and detention facilities where total evacuation is difficult to achieve.

- Voice-based systems provide response personnel with the ability to conduct orderly evacuation and notify building occupants of changing event circumstances.

- In high rise buildings, different evacuation messages may be played to each floor, depending on the location of the fire. The floor the fire is on along with ones above it may be told to evacuate while floors much lower may simply be asked to stand by.

Mass Notification Systems/Emergency Communication Systems

New codes and standards introduced around 2010 especially the new UL Standard 2572, the U.S. Department of Defense's UFC 4-021-01 Design and O&M Mass Notification Systems, and NFPA 72 2010 edition Chapter 24 have led fire alarm system manufacturers to expand their systems voice evacuation capabilities to support new requirements for mass notification including support for multiple types of emergency messaging (i.e. inclement weather emergency, security alerts, amber alerts). The major requirements of a mass notification system are to provide prioritized messaging according to the local facilities emergency response plan. The emergency response team must define the priority of potential emergency events at the site and the fire alarm system must be able to support the promotion and demotion of notifications based on this emergency response plan. Emergency Communication Systems also have requirements for visible notification in coordination with any audible notification activities to meet requirements of the Americans with Disabilities Act. Many manufacturers have made efforts to certify their equipment to meet these new and emerging standards. Mass notification system categories include the following:

- Tier 1 systems are in-building and provide the highest level of survivability.
- Tier 2 systems are out of the building and provide the middle level of survivability.
- Tier 3 systems are "At Your Side" and provide the lowest level of survivability.

Mass notification systems often extend the notification appliances of a standard fire alarm system to include PC based workstations, text-based digital signage, and a variety of remote notification options including email, text message, RSS feed , or IVR-based telephone text-to-speech messaging.

Building Safety Interfaces

Magnetic smoke door holders/retainers: Wall mounted solenoids or electromagnets controlled by a fire alarm system or detection component that magnetically secures spring-loaded self-closing smoke tight doors in the open position. Designed to de-magnetize to allow automatic closure of the door on command from the fire control or upon failure of the power source, interconnection or controlling element. Stored energy in the form of a spring or gravity can then close the door to restrict the passage of smoke from one space to another in an effort to maintain a tenable atmosphere on either side of the door during evacuation and fire fighting efforts in buildings. Electromagnetic fire door holders can be hard-wired into the fire panel, radio controlled triggered by radio waves from a central controller connected to a fire panel, or, acoustic, which learn the sound of the fire alarm and release the door upon hearing this exact sound. Duct mounted smoke detection: smoke detection mounted in such a manner as to sample the airflow through duct work and other plenums specifically fabricated for the transport of environmental air into conditioned spaces. Interconnection to the fan motor control circuits is intended to stop air movement, close dampers and generally prevent the recirculation of toxic smoke and fumes produced by fire into occupiable spaces.

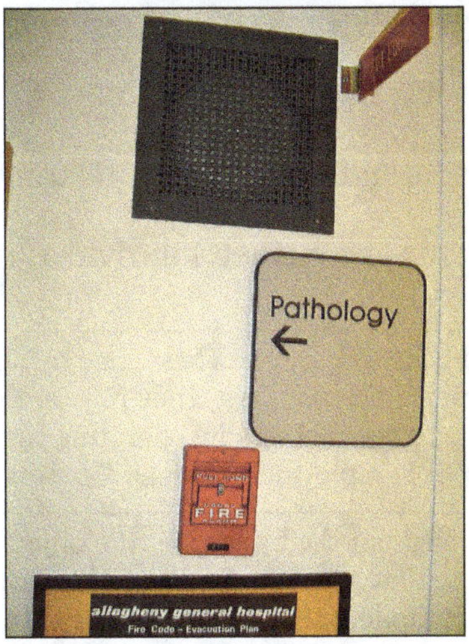

Coded fire alarm pull station below a 10" bell.

Emergency elevator service: Activation of automatic initiating devices associated with elevator operation are used to initiate emergency elevator functions, such as recall of associated elevator cab(s). The recall will cause the elevator cabs to return to the ground level for use by fire service response teams and to ensure that cabs do not return to the floor of fire incidence, in addition to prevent people from becoming trapped in the elevators. Phases of operation include primary recall (typically the ground level), alternate/secondary recall (typically a floor adjacent to the ground level – used when the initiation occurred on the primary level), illumination of the "fire hat" indicator when an alarm occurs in the elevator hoistway or associated control room, and in some cases shunt trip (disconnect) of elevator power (generally used where the control room or hoistway is protected by fire sprinklers).

Public address rack (PAR): An audio public address rack shall be interfaced with fire alarm system, by adding signaling control relay module to either rack power supply unit, or to the main amplifier driving this rack. The purpose is to "mute" the BGM (background music) of this rack in case of emergency in case of a fire initiating the true alarm.

US Fire Alarm System Categories

Fire alarm systems in non-domestic premises are generally designed and installed in accordance with the guidance given in BS 5839 Part 1. There are many types of fire alarm systems each suited to different building types and applications. A fire alarm system can vary dramatically in both price and complexity, from a single panel with a detector and sounder in a small commercial property to an addressable fire alarm system in a multi-occupancy building.

BS 5839 Part 1 categorizes fire alarm systems as:

- "M" manual system (no automatic fire detectors so the building is fitted with call points and sounders).
- "L" automatic systems intended for the protection of life.
- "P" automatic systems intended for the protection of property.

Categories for automatic systems are further subdivided into L1 to L5, and P1 to P2.

M	Manual systems, e.g. hand bells, gongs, etc. These may be purely manual or manual electric, the latter may have call points and sounders. They rely on the occupants of the building discovering the fire and acting to warn others by operating the system. Such systems form the basic requirement for places of employment with no sleeping risk.
P1	The system is installed throughout the building – the objective being to call the fire brigade as early as possible to ensure that any damage caused by the fire is minimized. Small low-risk areas can be excepted such as toilets and cupboards less than $1m^2$.
P2	Detection should be provided in parts of the building where the risk of ignition is high and the contents are particularly valuable. Category 2 systems provide fire detection in specified parts of the building where there is either high risk or where business disruption must be minimized.
L1	A category L1 system is designed for the protection of life and which has automatic detectors installed throughout all areas of the building (including roof spaces and voids) with the aim of providing the earliest possible warning. A category L1 system is likely to be appropriate for the majority of residential care premises. In practice, detectors should be placed in nearly all spaces and voids. With category 1 systems, the whole of a building is covered apart from minor exceptions.
L2	A category L2 system designed for the protection of life and which has automatic detectors installed in escape routes, rooms adjoining escape routes and high hazard rooms. In medium-sized premises (sleeping no more than ten residents), a category L2 system is ideal. These fire alarm systems are identical to an L3 system but with additional detection in an area where there is a high chance of ignition (e.g., kitchen) or where the risk to people is particularly increased (e.g., sleeping risk).
L3	This category is designed to give early warning to everyone. Detectors should be placed in all escape routes and all rooms that open onto escape routes. Category 3 systems provide more extensive cover than category 4. The objective is to warn the occupants of the building early enough to ensure that all are able to exit the building before escape routes become impassable.
L4	Category 4 systems cover escape routes and circulation areas only. Therefore, detectors will be placed in escape routes, although this may not be suitable depending on the risk assessment or if the size and complexity of a building is increased. Detectors might be sited in other areas of the building, but the objective is to protect the escape route.

L5	This is the "all other situations" category, e.g., computer rooms, which may be protected with an extinguishing system triggered by automatic detection. Category 5 systems are the "custom" category and relate to some special requirement that cannot be covered by any other category.

Zoning

An important consideration when designing fire alarms is that of individual zones. The following recommendations are found in BS 5839 Part 1:

- A single zone should not exceed 2,000m² in floor space.
- Where addressable systems are in place, two faults should not remove protection from an area greater than 10,000m².
- A building may be viewed as a single zone if the floor space is less than 300m².
- Where the floor space exceeds 300m² then all zones should be restricted to a single floor level.
- Stairwells, lift shafts or other vertical shafts (nonstop risers) within a single fire compartment should be considered as one or more separate zones.
- The maximum distance traveled within a zone to locate the fire should not exceed 60 m.

Also, the NFPA recommends placing a list for reference near the FACP showing the devices contained in each zone.

Emergency Population Warning

Horn for public warning system in Sweden.

An emergency population warning is a method whereby local, regional, or national authorities can contact members of the public en masse to warn them of an impending emergency. These warnings may be necessary for a number of reasons, including:

- Weather emergencies such as tornadoes, hurricanes, and ice storms.
- Geological disasters such as earthquakes, landslides, volcanic eruptions, and tsunamis.
- Industrial Disasters Such As The Release Of Toxic Gas Or Contamination Of River Water.
- Radiological disasters such as a nuclear plant disaster.
- Medical emergencies such as an outbreak of a fast-moving infectious disease.
- Warfare or acts of terrorism.

Many local areas use emergency population warnings to advise of prison escapes, abducted children, emergency telephone number outages, and other events.

Requirements

In order to develop an effective emergency warning system, certain things are required:

- An agreement as to what constitutes an emergency in the area served by the system. This differs from region to region depending on the local climate, geology, and the like.
- An agreement as to who can initiate an alert. In some countries all warnings are transmitted by a single command center, while in others (such as the United States) a host of local, regional, and national agencies are authorized to initiate warnings.
- A system or systems by which the information can be quickly transmitted to the population.
- An education program to teach the general public how to recognize an alert or what to do if a warning is broadcast.

Emergency Alert System

The Emergency Alert System (EAS) is a national warning system in the United States put into place on January 1, 1997 (approved by the Federal Communications Commission), when it replaced the Emergency Broadcast System (EBS), which in turn replaced

CONELRAD. It is jointly coordinated by the Federal Emergency Management Agency (FEMA), the Federal Communications Commission (FCC) and the National Oceanic and Atmospheric Administration (NOAA). The EAS regulations and standards are governed by the Public Safety and Homeland Security Bureau of the FCC.

Current logo.

As with its predecessors, the system is primarily designed to allow the president to address the country via all radio and television stations, in the event of a national emergency. Despite this, neither the system nor its predecessors have been used in this manner, due to the ubiquity of news coverage in these situations. In practice, it is more commonly used to distribute information regarding imminent threats to public safety, such as severe weather situations (including flash floods and tornadoes), AMBER Alerts of child abductions, and other civic emergencies.

Authorized organizations are able to disseminate and coordinate emergency alerts and warning messages through EAS. The Integrated Public Alert and Warning System (IPWAS) is used as a backend to distribute alert information via EAS and related technologies such as Wireless Emergency Alerts (WEA), using Common Alerting Protocol (CAP). EAS messages are transmitted primarily via terrestrial and satellite radio and television (including broadcast and multichannel television), which are required to participate in the system.

Technical Concept

Messages in the EAS are composed of four parts: a digitally encoded Specific Area Message Encoding (SAME) header, an attention signal, an audio announcement, and a digitally encoded end-of-message marker.

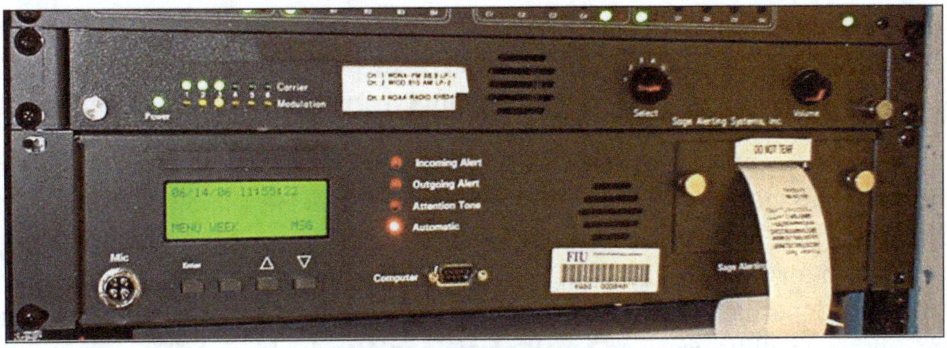

A Sage EAS ENDEC unit.

The SAME header (help·info) is the most critical part of the EAS design. It contains information about who originated the alert (the president, state or local authorities, the National Weather Service (NOAA/NWS), or the broadcaster), a short, general description of the event (tornado, flood, severe thunderstorm), the areas affected (up to 32 counties or states), the expected duration of the event (in minutes), the date and time it was issued (in UTC), and an identification of the originating station (see SAME for a complete breakdown of the header).

There are 77 radio stations designated as National Primary Stations in the Primary Entry Point (PEP) System to distribute presidential messages to other broadcast stations and cable systems.

The Emergency Action Notification is the notice to broadcasters that the president of the United States or their designee will deliver a message over the EAS via the PEP system. The government has stated that the system would allow a president to speak during a national emergency within 10 minutes.

Primary Entry Point Stations

The National Public Warning System, also known as the Primary Entry Point (PEP) stations, are a network of 77 radio stations that are, in coordination with FEMA, used to originate emergency alert and warning information to the public before, during, and after incidents and disasters. PEP stations are equipped with additional and backup communications equipment and power generators designed to enable them to continue broadcasting information to the public during and after an event. Beginning with WJR/Detroit and WLW/Cincinnati in 2016, FEMA began the process of constructing transportable studio shelters at the transmitters of 33 PEP stations, which feature broadcasting equipment, emergency provisions, a rest area, and an air filtration system. NPWS project manager Manny Centeno explained that these shelters were designed to "expand the survivability of these stations to include an all hazards platform, which means chemical, biological, radiological air protection and protection from electromagnetic pulse."

Communication Links

The FEMA National Radio System (FNARS) "Provides Primary Entry Point service to the Emergency Alert System", and acts as an emergency presidential link into the EAS. The FNARS net control station is located at the Mount Weather Emergency Operations Center.

Once an EAN is received by an EAS participant from a PEP station (or any other participant) the message then "daisy chains'" through the network of participants. "Daisy chains" form when one station receives a message from multiple other stations and the station then forwards that message to multiple other stations. This process creates many redundant paths through which the message may flow increasing the likelihood that the message will be received by all participants and adding to the survivability of the system. Each EAS participant is required to monitor at least two other participants.

EAS Header

Because the header lacks error detection codes, it is repeated 3 times for redundancy. However, the repetition of the data can itself be considered an error detection and correction code – like any error detection or correction code, it adds redundant information to the signal in order to make errors identifiable. EAS decoders compare the received headers against one another, looking for an exact match between any two, eliminating most errors which can cause an activation to fail. The decoder then decides whether to ignore the message or to relay it on the air if the message applies to the local area served by the station (following parameters set by the broadcaster).

The SAME header bursts are followed by an attention tone, which lasts between 8 and 25 seconds, depending on the originating station. The tone is 1050 Hz (help·info) on a NOAA Weather Radio station. On commercial broadcast stations, a "two-tone" (help·info) attention signal of 853 Hz and 960 Hz sine waves is used instead, the same signal used by the older Emergency Broadcast System. These tones have become infamous, and can be considered both frightening and annoying by viewers; in fact, the two tones, which form approximately the interval of a just major second at an unusually high pitch, were chosen specifically for their ability to draw attention, due to their unpleasantness on the human ear. The SAME header is equally known for its shrillness, which many have found to be startling. The "two-tone" system is no longer required as of 1998, and is to be used only for audio alerts before EAS messages. Like the EBS, the attention signal is followed by a voice message describing the details of the alert.

The message ends with 3 bursts of the AFSK "EOM", or End of Message, which is the text NNNN, preceded each time by the binary 10101011 calibration.

IPAWS

Under a 2006 executive order issued by George W. Bush, the U.S. government was instructed to create "an effective, reliable, integrated, flexible, and comprehensive" public warning system. This was accomplished via expansions to the aforementioned PEP network, and the development of the Integrated Public Alert and Warning System (IPAWS)—a national aggregator and distributor of alert information using the XML-based Common Alerting Protocol (CAP) and an internet network. IPAWS can be used to distribute alert information to EAS participants, supported mobile phones (Wireless Emergency Alerts), and other platforms.

Under an FCC report and order issued in 2007, EAS participants would be required to migrate to digital equipment supporting CAP within 180 days of the specification's adoption by FEMA. This officially occurred September 30, 2010, but the deadline was later delayed to June 30, 2012 at the request of broadcasters.

As part of the implementation, the FCC mandated that visual displays of alerts be able to display any extended text contained within the associated CAP message, when possible. The FCC has established that IPAWS is not a full substitute for the existing SAME protocol, as it is vulnerable to situations that may make internet connectivity unavailable. Therefore, broadcasters must convert CAP messages to legacy SAME headers to enable backwards compatibility with the existing "daisy chain" method of EAS distribution, providing a backup distribution path.

Station Requirements

The FCC requires all broadcast stations and multichannel video programming distributors (MVPD) to install and maintain FCC-certified EAS decoders and encoders at their control points or headends. These decoders continuously monitor the signals from other nearby broadcast stations for EAS messages. For reliability, at least two source stations must be monitored, one of which must be a designated *local primary*. Stations are to retain the latest version of the EAS handbook.

Stations are required by federal law to keep logs of all received messages. Logs may be kept by hand but are usually kept automatically by a small receipt printer in the encoder/decoder unit. Logs may also be kept electronically inside the unit as long as there is access to an external printer or method to transfer them to a personal computer.

In addition to the audio messages transmitted by radio stations, "television broadcast stations shall transmit a visual message containing the Originator, Event, Location and the valid time period of an EAS message". This may be a text "crawl" or a static visual message. A text "crawl" is displayed at the on-screen that contains all of the information encoded in the initial SAME header, as well as any extended text.that is contained within the associated CAP message. A color-coded "crawl" system is often used where the color signifies the priority of the message. Some television stations transmit only a static *slide* containing the required information. A television station may be used for monitoring by another station and thus the audio is necessary.

Stations are required by federal law to relay Emergency Action Notification (EAN) messages immediately (47 CFR Part 11.54). Stations traditionally have been allowed to opt out of relaying other alerts such as severe weather, and child abduction emergencies (AMBER Alerts) if they so choose (in practice, severe weather alerts are typically delegated to a station's local news department, and presented on-air with analysis by a staff meteorologist).

Also, many radio stations in your area may broadcast information about the emergency, be ready to hear this noise of emergency.

System Tests

All EAS equipment must be tested on a weekly basis. The required weekly test (RWT) consists, at a minimum, of the header and end-of-message tones. Though an RWT does not need an audio or graphic message announcing the test, many stations provide them as a courtesy to the public. In addition, television stations are not required to transmit a video message for weekly tests. RWTs are scheduled by the station on random days and times, (though quite often during late night or early afternoon hours), and are generally not relayed.

Required monthly tests (RMTs) are generally originated by the local or state primary station, a state emergency management agency, or by the National Weather Service and are then relayed by broadcast stations and cable channels. RMTs must be performed between 8:30 a.m. and local sunset during odd numbered months, and between local sunset and 8:30 a.m. during even numbered months. Received monthly tests must be retransmitted within 60 minutes of receipt. Additionally, an RMT should not be scheduled or conducted during an event of great importance such as a pre-announced presidential speech, coverage of a national/local election, major local or national news coverage outside regularly scheduled newscast hours or a major national sporting event such as the Super Bowl or World Series, with other events such as the Indianapolis 500 and Olympic Games mentioned in individual EAS state plans.

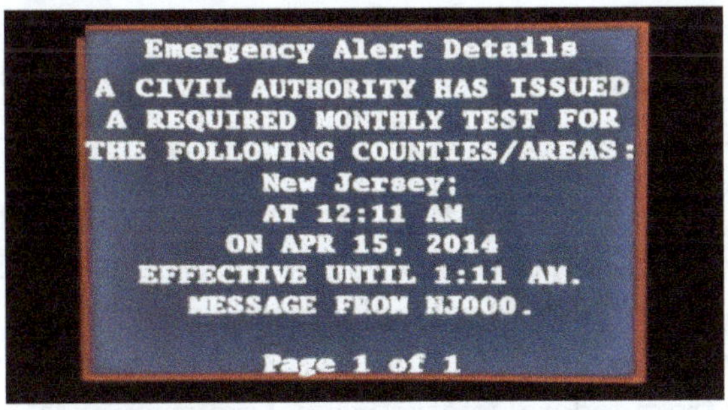

A Required Monthly Test (RMT) transmitted in New Jersey on April 15, 2014 as shown on a television set.

An RWT is not required during a calendar week in which an RMT is scheduled. No testing has to be done during a calendar week in which all parts of the EAS (header burst, attention signal, audio message, and end of message burst) have been legitimately activated.

In July 2018, in response to the aftermath of the false missile alert in Hawaii earlier in the year (which was caused by operator error during an internal drill protocol), the FCC announced that it would take steps to promote public awareness and improve efficiency of the system, including requiring safeguards to prevent distribution of false alarms, the ability to authorize "live code" tests—which would simulate the process and response to an actual emergency, and authorizations to use the EAS tones in public service announcements that promote awareness of the system.

Additions and Proposals

The number of event types in the national system has grown to eighty. At first, all but three of the events (civil emergency message, immediate evacuation, and emergency action notification "national emergency") were weather-related (such as a tornado warning). Since then, several classes of non-weather emergencies have been added, including, in most states, the AMBER Alert System for child abduction emergencies. In 2016, three additional weather alert codes were authorized for use in relation to hurricane events, including Extreme Wind Warning (EWW), Storm Surge Warning (SSW) and Storm Surge Watch (SSA).

In 2004, the FCC issued a Notice of Proposed Rulemaking (NPR) seeking comment on whether EAS in its present form is the most effective mechanism for warning the American public of an emergency and, if not, on how EAS can be improved, such as mandatory text messages to cellphones, regardless of subscription. As noted above, rules implemented by the FCC on July 12, 2007 provisionally endorse incorporating CAP with the SAME protocol.

In 2018, Hawaii Senator Brian Schatz proposed the Reliable Emergency Alert Distribution Improvement (READI) Act; alongside proposing that FEMA create best practices for EAS usage and false alarms (the latter in response to the Hawaii false missile alert), it would have required that certain types of alerts receive repeat broadcasts in order to ensure that viewers receive the alert.

EAS for Consumers

The EAS is designed to be useful for the entire public, not just those with SAME-capable equipment. However, several consumer-level radios do exist, especially weather radio receivers, which are available to the public through both mail-order and retailers. Other specialty receivers for AM/FM/ACSSB (LM) are available only through mail-order, or in some places from federal, state, or local governments, especially where there is a potential hazard nearby such as a chemical factory. These radios come pre-tuned to a station in each area that has agreed to provide this service to local emergency management officials and agencies, often with a direct link back to the plant's safety system or control room for instant activation should an evacuation or other emergency arise.

The ability to narrow messages down so that only the actual area in danger is alerted is extremely helpful in preventing false warnings, which was previously a major tune-out factor. Instead of sounding for all warnings within a station's area, SAME-decoder radios now sound only for the counties for which they are programmed. When the alarm sounds, anyone with the radio *knows* that the danger is nearby and protective action should be taken. For this reason, the goal of the National Weather Service is that each home should have both a smoke detector and a SAME weather radio.

Limitations

The EAS can only be used to relay audio messages that preempt all programming; as the intent of an Emergency Action Notification is to serve as a "last-ditch effort to get a message out if the president cannot get to the media", it can easily be made redundant by the immediate and constant coverage that major weather events and other news-worthy situations—such as, most prominently, the September 11 attacks in 2001—receive from television broadcasters and news channels. Following the attacks, then-FCC chairman Michael K. Powell cited "the ubiquitous media environment" as justification for not using the EAS in their immediate aftermath. Glenn Collins of *The New York Times* acknowledged these limitations, noting that "no president has ever used the current (EAS) system or its technical predecessors in the last 50 years, despite the Soviet missile crisis, a presidential assassination, the Oklahoma City bombing, major earthquakes and three recent high-alert terrorist warnings", and that using it would have actually hindered the availability of live coverage from media outlets.

Following the tornado outbreak of March 3, 2019, Birmingham, Alabama NWS meteorologist Kevin Laws told CNN that he, personally, wished that alerts could be updated in real-time in order to reflect the unpredictable nature of weather events –noting that the storm system's unexpected change in trajectory towards Lee County resulted in only a nine-minute warning (the resulting tornado would kill 23 people).

The trend of cord cutting has led to concerns that viewers' lessened use of broadcast media in favor of streaming video services would inhibit their ability to receive emergency information (notwithstanding availability of alerts on mobile phones).

Incidents

Cybersecurity Breaches

On several notable occasions, EAS equipment became the subject of hacks by outside entities due to poor security measures, including poor firewalls, and use of insecure or unchanged default passwords on encoder hardware. In February 2013, several stations in Great Falls, Montana and Marquette, Michigan were breached in such a manner, relaying a hoax message warning that "dead bodies" were "rising from their graves". It was determined that the stations' equipment was poorly protected, and that the broadcasters had further neglected to change default factory logins or passwords, opting to use factory presets instead. Because of this, the FCC, FEMA, equipment manufacturers, as well as trade groups, including the Michigan Association of Broadcasters, urged broadcasters to change their passwords and to recheck their security measures.

In a related incident, WIZM-FM in La Crosse, Wisconsin accidentally triggered the EAS on television station WKBT-DT by airing a recording of the false message during its morning show. The hosts' laughter was heard in the audio relayed.

On February 28, 2017, WZZY in Winchester, Indiana was hacked in a nearly-identical manner, playing the same "dead bodies" audio from the February 2013 incidents. The incident prompted a public response from the Randolph County Sheriff's Department clarifying that there was no actual emergency.

Tone usage Outside of Alerts

To protect the integrity of the system, and prevent false activations, the FCC prohibits the use of actual or simulated EAS/WEA tones and attention signals outside of genuine alerts, tests, or authorized public service announcements, especially when they are used "to capture audience attention during advertisements; dramatic, entertainment, and educational programs". Broadcasters who misuse the tones may be sanctioned (including being required to partake in compliance measures) and fined.

- Tones from the EAS were used in the trailer for the 2013 film Olympus Has Fallen; cable providers were fined $1.9 million by the Federal Communications Commission (FCC) on March 3, 2014 for misuse of EAS tones. An event similar to this previously occurred in November 2013, when TBS was fined $25,000 for the use of EAS tones in a Conan advertisement.

- During the October 24, 2014 episode of Bobby Bones' nationally-syndicated radio show, Bones played audio from the 2011 national test as part of a rant about a genuine test from Nashville's Fox affiliate, WZTV, that interrupted Game 2 of the 2014 World Series on October 22. The errant Emergency Action Notification was relayed to some broadcasters and cable systems—particularly those not configured to reject EAN messages that did not match the current date. On May 19, 2015, iHeartMedia (who distributes the show and owns its flagship station, WSIX-FM) was fined $1 million for the incident, and was ordered to implement a three-year compliance plan and remove all EAS tones or similar-sounding noises from its audio production libraries in order to avoid any further incidents.

- From August 4 to August 6, 2016, Tegna, Inc.-owned NBC affiliate WTLV in Jacksonville, Florida aired an ad several times during NBC's primetime coverage of the 2016 Summer Olympics produced by the marketing department of the National Football League's Jacksonville Jaguars featuring out-of-sequence EAS tones over Jaguars training camp footage and a voiceover noting "this is not a test, this is an emergency broadcast transmission seek shelter immediately", along with the on-screen text "the storm is coming". The ad aired four times before station compliance authorities pulled the advertisement after the local news industry blog FTVLive criticized the station for carrying it, especially during the peak of the Atlantic hurricane season. FTVLive's piece would be noted by the FCC in their decision against WTLV rendered on May 30, 2017, when it was given a $55,000 fine for carrying the offending Jaguars ad.

- The FCC issued several fines relating to EAS tone usage in August 2019, including ABC being fined $395,000 for using wireless emergency alert tones multiple times during a Jimmy Kimmel Live sketch, AMC Networks being fined $104,000 for using the tones in The Walking Dead episode "Omega", Discovery Inc. being fined $68,000 for including footage of an actual WEA activation during a Lone Star Law episode that was set during Hurricane Harvey, and Meruelo Group was fined $61,000 for including an EAS-like tone during a radio advertisement for KDAY and KDEY-FM's morning show.

Testing Errors

- On October 19, 2008, KWVE-FM in San Clemente, California accidentally initiated a Required Monthly Test when it meant to conduct a Required Weekly Test. Furthermore, an operator aborted the test mid-way through the broadcast (failing to broadcast the end-of-message tone), causing all area outlets to broadcast KWVE-FM's programming until those stations took their equipment offline. On September 15, 2009, the FCC fined the station's owner, Calvary Chapel Costa Mesa, $5,000. After the fine was levied, various state broadcast associations in the United States submitted joint letters to the FCC, protesting against the fine, saying that the commission could have handled the matter better. On November 13, 2009, the FCC rescinded its fine against KWVE-FM, but had still admonished the station for broadcasting an unauthorized RMT, as well as omitting the code to end the test.

- On September 21, 2017, a technical glitch in another scheduled test by KWVE caused the end-of-message tone to be omitted, causing regional participants (particularly Charter and Cox Cable systems in Orange County) to broadcast a portion of Chuck Swindoll's Insight for Living program from the station. In the audio, Swindoll was heard quoting 2 Timothy 3:1 from the Bible, stating that "extremely violent times will come". Due to the nature of the quotes (which could be insinuated out of context to be apocalyptic in nature), some viewers speculated that the accident was the result of a hack.

Tsunami Warning System

A tsunami warning system (TWS) is used to detect tsunamis in advance and issue warnings to prevent loss of life and damage to property. It is made up of two equally important components: a network of sensors to detect tsunamis and a communications infrastructure to issue timely alarms to permit evacuation of the coastal areas. There are two distinct types of tsunami warning systems: international and regional. When operating, seismic alerts are used to instigate the watches and warnings; then, data from observed sea level height (either shore-based tide gauges or DART buoys) are used to verify the existence of a tsunami. Other systems have been proposed to augment the warning procedures; for example, it has been suggested that the duration and frequency content of t-wave energy (which is earthquake energy trapped in the ocean SOFAR channel) is indicative of an earthquake's tsunami potential.

The first rudimentary system to alert communities of an impending tsunami was attempted in Hawaii in the 1920s. More advanced systems were developed in the wake of the April 1, 1946 (caused by the Aleutian Islands earthquake) and May 23, 1960 (caused by the Valdivia earthquake) tsunamis which caused massive devastation in Hilo, Hawaii. While tsunamis travel at between 500 and 1,000 km/h (around 0.14 and 0.28 km/s) in open water, earthquakes can be detected almost at once as seismic waves travel with a typical speed of 4 km/s (around 14,400 km/h). This gives time for a possible tsunami forecast to be made and warnings to be issued to threatened areas, if warranted. Until a reliable model is able to predict which earthquakes will produce significant tsunamis, this approach will produce many more false alarms than verified warnings.

Evacuation route sign in a low-lying coastal area on the West Coast of the United States.

International Systems (IS)

Pacific Ocean

Tsunami warnings for most of the Pacific Ocean are issued by the Pacific Tsunami Warning Center (PTWC), operated by the United States NOAA in Ewa Beach, Hawaii. NOAA's National Tsunami Warning Center (NTWC) in Palmer, Alaska issues warnings for North America, including Alaska, British Columbia, Oregon, California, the Gulf of Mexico, and the East coast. The PTWC was established in 1949, following the 1946 Aleutian Island earthquake and a tsunami that resulted in 165 casualties on Hawaii and in Alaska; NTWC was founded in 1967. International coordination is achieved through the International Coordination Group for the Tsunami Warning System in the Pacific, established by the Intergovernmental Oceanographic Commission of UNESCO.

Japanese Tsunami warning sign.

In 2017, The US congressional panel passed a vote to continue funding a global tsunami detection system that gives U.S. officials an accurate forecast in order to decrease damage cause by Tsunamis.

Indian Ocean (ICG/IOTWMS)

After the 2004 Indian Ocean Tsunami which killed almost 250,000 people, a United Nations conference was held in January 2005 in Kobe, Japan, and decided that as an initial step towards an International Early Warning Programme, the UN should establish an Indian Ocean Tsunami Warning System. This resulted in a warning system for Indonesia and other affected areas. Indonesia's system fell out of service in 2012 because the detection buoys were no longer operational. Tsunami prediction is therefore currently limited to detection of seismic activity; there is no system to predict tsunamis based on volcanic eruptions.

Tsunami Early Warning Tower board in Hikkaduwa, Sri Lanka.

Indonesia was hit by tsunamis in September and December 2018. The December 2018 tsunami was caused by a volcano.

North Eastern Atlantic, the Mediterranean and Connected Seas (ICG/NEAMTWS)

The First United Session of the Inter-governmental Coordination Group for the Tsunami Early Warning and Mitigation System in the North Eastern Atlantic, the Mediterranean and connected Seas (ICG/NEAMTWS), established by the Intergovernmental Oceanographic Commission of UNESCO Assembly during its 23rd Session in June 2005, through Resolution XXIII.14, took place in Rome on 21 and 22 November 2005.

The meeting, hosted by the Government of Italy (the Italian Ministry of Foreign Affairs and the Italian Ministry for the Environment and Protection of Land and Sea), was attended by more than 150 participants from 24 countries, 13 organizations and numerous observers.

Caribbean

A Caribbean-wide tsunami warning system was planned to be instituted by the year 2010, by representatives of Caribbean nations who met in Panama City in March 2008. Panama's last major tsunami killed 4,500 people in 1882. Barbados has said it will review or test its tsunami protocol in February 2010 as a regional pilot.

Regional Warning Systems

Regional (or local) warning system centres use seismic data about nearby recent earthquakes to determine if there is a possible local threat of a tsunami. Such systems are capable of issuing warnings to the general public (via public address systems and sirens) in less than 15 minutes. Although the epicenter and moment magnitude of an underwater quake and the probable tsunami arrival times can be quickly calculated, it is almost always impossible to know whether underwater ground shifts have occurred which will result in tsunami waves. As a result, false alarms can occur with these systems, but the disruption is small, which makes sense due to the highly localised nature of these extremely quick warnings, in combination with how difficult it would be for a false alarm to affect more than a small area of the system. Real tsunamis would affect more than just a small portion.

Tsunami warning system in East Timor.

Japan

Japan has a nationwide tsunami warning system. The system usually issues the warning minutes after an Earthquake Early Warning (EEW) is issued, should there be expected waves. The tsunami warning was issued within 3 minutes with the most serious rating on its warning scale during the 2011 Tōhoku earthquake and tsunami; it was rated as a "major tsunami", being at least 3 m (9.8 ft) high. An improved system was unveiled on March 7, 2013 following the 2011 disaster to better access the waves.

Conveying the Warning

Detection and prediction of tsunamis is only half the work of the system. Of equal importance is the ability to warn the populations of the areas that will be affected. All tsunami warning systems feature multiple lines of communications (such as SMS, e-mail, fax, radio, texting and telex, often using hardened dedicated systems) enabling emergency messages to be sent to the emergency services and armed forces, as well to population-alerting systems (e.g. sirens) and systems like the Emergency Alert System.

Mast with warning system, and sign detailing escape routes, on the coast of Okumatsushima, Miyagi prefecture, Japan.

Shortcomings

With the speed at which tsunami waves travel through open water, no system can protect against a very sudden tsunami, where the coast in question is too close to the epicenter. A devastating tsunami occurred off the coast of Hokkaidō in Japan as a result of an earthquake on July 12, 1993. As a result, 202 people on the small island of Okushiri, Hokkaido lost their lives, and hundreds more were missing or injured. This tsunami struck just three to five minutes after the quake, and most victims were caught while fleeing for higher ground and secure places after surviving the earthquake. This was also the case in Aceh, Indonesia.

While there remains the potential for sudden devastation from a tsunami, warning systems can be effective. For example, if there were a very large subduction zone earthquake (moment magnitude 9.0) off the west coast of the United States, people in Japan, would therefore have more than 12 hours (and likely warnings from warning systems in Hawaii and elsewhere) before any tsunami arrived, giving them some time to evacuate areas likely to be affected.

Earthquake Warning System

An earthquake warning system is a system of accelerometers, seismometers, communication, computers, and alarms that is devised for notifying adjoining regions of a substantial earthquake while it is in progress. This is not the same as earthquake prediction, which is currently incapable of producing decisive event warnings.

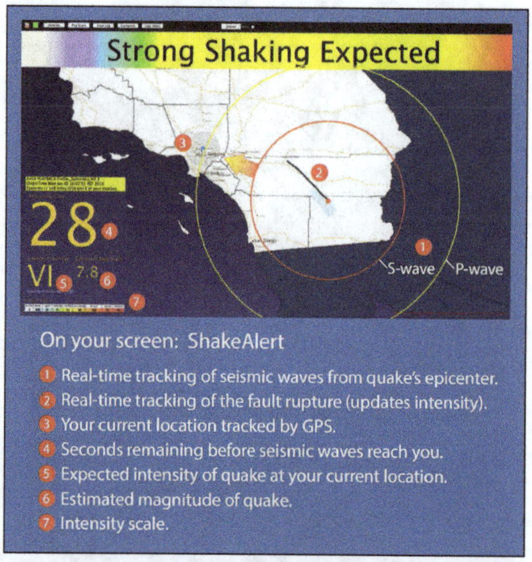

Example of early warning issued by ShakeAlert.

Time Lag and Wave Projection

An earthquake is caused by the release of stored elastic strain energy during rapid sliding along a fault. The sliding starts at some location and progress away from the hypocenter in each direction along the fault surface. The speed of the progression of this fault tear is slower than, and distinct from the speed of the resultant pressure and shear waves, with the pressure wave traveling faster than the shear wave. The pressure waves generate an abrupt shock. The shear waves generate periodic motion (at about 1 Hz) that is the most destructive to structures, particularly buildings that have a similar resonant period. Typically these buildings are around eight floors in height. These waves will be strongest at the ends of the slippage, and may project destructive waves well beyond the fault failure. The intensity of such remote effects are highly dependent upon local soils conditions within the region and these effects are considered in constructing a model of the region that determines appropriate responses to specific events.

Transit Safety

Such systems are currently implemented to determine appropriate real-time response to an event by the train operator in urban rail systems such as BART (Bay Area Rapid Transit). The appropriate response is dependent on the warning time, the local right-of-way conditions and the current speed of the train.

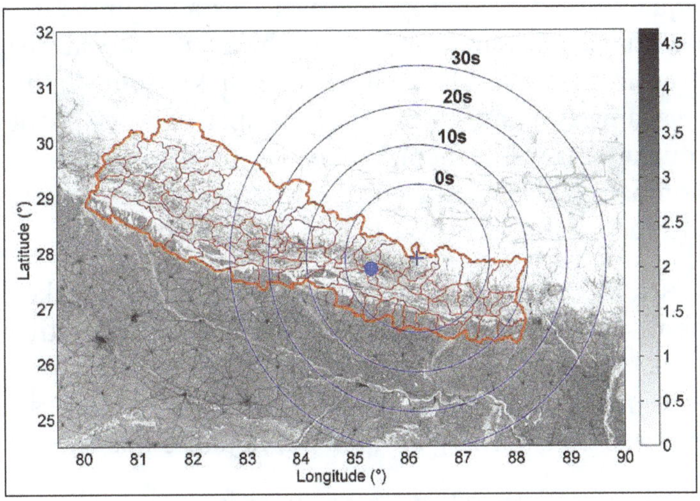

Warning time given by the earthquake warning system of the Earthquake Network project during the May 2015 Nepal earthquake. The cross marker depicts the earthquake epicenter while the dot marker shows the detection location.

Deployment

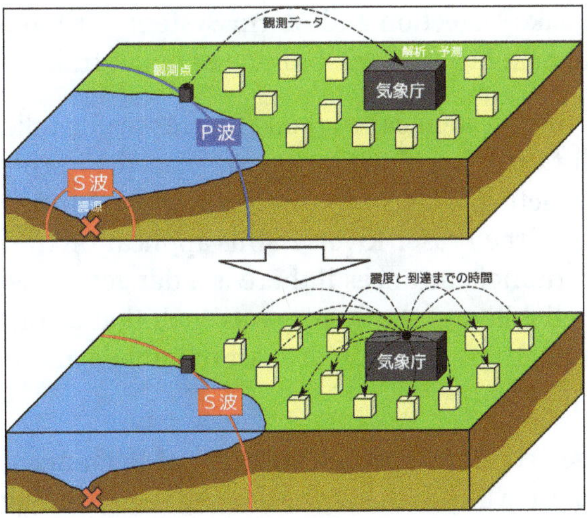

The Earthquake Early Warning in Japan: When P-waves are detected, the readings are analyzed immediately and the warning information is distributed to advanced users such as; broadcasting stations and mobile phone companies, before the arrival of S-waves (lower).

As of 2016, Japan and Taiwan have comprehensive, nationwide earthquake early warning systems. Other countries and regions have limited deployment of earthquake warning systems, including Mexico (the Mexican Seismic Alert System covers areas of central and southern Mexico including Mexico City and Oaxaca), limited regions of Romania (the Basarab bridge in Bucharest), and parts of the United States. The earliest automated earthquake pre-detection systems were installed in the 1990s; for instance, in California, the Calistoga fire station's system which automatically triggers a citywide siren to alert the entire area's residents of an earthquake. Some California fire departments use their warning systems to automatically open overhead doors of fire stations before the earthquake can disable them. While many of these efforts are governmental, several private companies also manufacture earthquake early warning systems to protect infrastructure such as elevators, gas lines and fire stations.

In the 2019 July 4th M6.4 and July 5th M7.1 Ridgecrest earthquakes, Early Warning Labs became the first and only commercially licensed USGS partner to successfully deliver ShakeAlerts to people & automated systems across Southern California. Early Warning Labs consumer app QuakeAlert, was the only app to successfully deliver mobile alerts. The average user warning for both earthquakes was 45 seconds.

The Mexican Seismic Alert System began operations in 1991 and began publicly issuing alerts in 1993. Initially serving Mexico City with twelve sensors, the system now has 97 sensors and is designed to protect life and property in several central and southern Mexican states.

Japan's Earthquake Early Warning system was put to practical use in 2006. The system that warns the general public was installed on October 1, 2007. It was modeled partly on the Urgent Earthquake Detection and Alarm System (UrEDAS) of Japan Railways, which was designed to enable automatic braking of bullet trains.

In 2009, an early warning system called ShakeAlarm was installed and commissioned in Vancouver, British Columbia, Canada. It was placed to protect a piece of critical transportation infrastructure called the George Massey Tunnel, which connects the north and south banks of the Fraser River. In this application the system automatically closes the gates at the tunnel entrances if there is a dangerous seismic event inbound. The success and the reliability of the system was such that as of 2015 there have been several additional installations on the west coast of Canada and the United States, and there are more being planned.

In January 2013, Francesco Finazzi of the University of Bergamo started the Earthquake Network research project which aims at developing and maintaining a crowdsourced earthquake warning system based on smartphone networks. Smartphones are used to detect the ground shaking induced by an earthquake and a warning is issued as soon as an earthquake is detected. People living at a further distance from the epicenter and the detection point may be alerted before they are reached by the damaging waves of the earthquake. People can take part in the project by installing the Android application "Earthquake Network" on their smart phones. The app requires the phone to receive the alerts.

In December 2014, the United States Congress approved a $5 million allocation as part of the Consolidated Appropriations Act, 2014 in order to expand funding for development of the system.

The United States Geological Survey (USGS) has investigated collaboration with the social networking site Twitter to allow for more rapid construction of ShakeMaps. As of May 2015, USGS began testing an earthquake early warning app. QuakeAlert was developed by Early Warning Labs in official partnership with USGS.

In July 2015, the USGS awarded $4 million in funding to the Berkeley Seismological Laboratory, the California Institute of Technology and the University of Washington, to turn the current ShakeAlert early warning prototype into a more robust system that could be used by "cities, industries, utilities and transportation networks in California, Oregon and Washington".

Gravimetric data from the 2011 Tōhoku earthquake has been used to create a model for increased warning time compared to seismic models, as gravity fields travel at the speed of light, much faster than seismic waves.

Gale Warning

Weather services issue a gale warning for maritime locations currently or imminently experiencing winds of gale force on the Beaufort scale. Gale warnings allow mariners to take precautionary actions to ensure their safety at sea or to seek safe anchorage and ride out the storm on land. Though usually associated with deep low-pressure areas, winds strong enough to catalyze a gale warning can occur in other conditions too, including from anticyclones, or high-pressure systems, in the continental interior.

6

Emergency Management Software

There are many emergency management software and applications that are widely used. This chapter sheds light on these emergency management software to provide an in-depth understanding of the subject.

Emergency management software provides businesses, organizations, and emergency response professionals the tools to prevent, prepare for, and manage a variety of emergencies or incidents in the workplace. Any emergency management personnel or agency can utilize this type of software to simulate possible emergency scenarios, operationalize procedures, and coordinate notifications and messaging. Emergency management software helps develop business continuity plans and adapt them according to post-crisis feedback and analysis, ensuring organizations quickly and effectively return to normal operations and levels of productivity.

Emergency management solutions are more comprehensive than the mass notification and alerting functionalities of emergency notification software solutions. Emergency management software differs also from incident management software products, the latter of which provides IT teams the tools to proactively investigate, respond, and curb IT emergencies, rather than public safety emergencies.

To qualify for inclusion in the Emergency Management category, a product must:

- Configure to the mission-critical and operational needs of a variety of industries.
- Gather and leverage data from tools such as GIS, weather channels, or disaster status systems to power emergency plans and strategies.
- Communicate with and relay information to industries and departments such as hospitals, public health, and state governments.
- Analyze the reactions to implemented emergency plans for future "what if" scenarios.

Real-time Emergency Management Applications

The last years have been characterized by the arising of highly distributed computing platforms composed of a heterogeneity of computing and communication resources including centralized high-performance computing architectures (e.g. clusters or large shared-memory machines), as well as multi-/many-core components also integrated into mobile nodes and network facilities. The emerging of computational paradigms such as Grid and Cloud Computing, provides potential solutions to integrate such platforms with data systems, natural phenomena simulations, knowledge discovery and decision support systems responding to a dynamic demand of remote computing and communication resources and services.

In this context time-critical applications, notably emergency management systems, are composed of complex sets of application components specialized for executing specific computations, which are able to cooperate in such a way as to perform a global goal in a distributed manner. Since the last years the scientific community has been involved in facing with the programming issues of distributed systems, aimed at the definition of applications featuring an increasing complexity in the number of distributed components, in the spatial distribution and cooperation between interested parties and in their degree of heterogeneity.

Over the last decade the research trend in distributed computing has been focused on a crucial objective. The wide-ranging composition of distributed platforms in terms of different classes of computing nodes and network technologies, the strong diffusion of applications that require real-time elaborations and online compute-intensive processing as in the case of emergency management systems, lead to a pronounced tendency of systems towards properties like self-managing, self-organization, self-controlling and strictly speaking adaptivity.

Adaptivity implies the development, deployment, execution and management of applications that, in general, are dynamic in nature. Dynamicity concerns the number and the specific identification of cooperating components, the deployment and composition of the most suitable versions of software components on processing and networking resources and services, i.e., both the quantity and the quality of the application components to achieve the needed Quality of Service (QoS). In time-critical applications the QoS specification can dynamically vary during the execution, according to the user intentions and the information produced by sensors and services, as well as according to the monitored state and performance of networks and nodes.

The general reference point for this kind of systems is the Grid paradigm which, by definition, aims to enable the access, selection and aggregation of a variety of distributed and heterogeneous resources and services. However, though notable advancements have been achieved in recent years, current Grid technology is not yet able to supply the needed software tools with the features of high adaptivity, ubiquity, proactivity, self-organization, scalability and performance, interoperability, as well as fault tolerance and security, of the emerging applications.

Related Works

When a distributed application is executed on a dynamic execution environment, where the number and features of nodes and network facilities but also the desired QoS level may vary significantly and in unpredictable ways, we need to design systems which are able to modify their actual configuration for maintaining and respecting the desired objectives. For application configuration we intend a specific identification of application components, their mapping onto the available computing resources, the specification of their internal behaviors (e.g. if they exploit a sequential or a parallel elaboration) and the way in which the computations are performed (i.e. for a sequential component the selected algorithm whereas for a parallel computation the exploited parallelism scheme). The actual system configuration can be modified by exploiting dynamic reconfiguration activities classified in four general categories.

- Geometrical changes: Affect the mapping between the internal structure of an application component and the system resources on which it is currently executed. Such class of changes consist in migrating a component in response to specific conditions, e.g. when a system resource fails or a new node is included in the system. These reconfigurations can be useful for load-balancing or for improving the service reliability.
- Structural changes: Affect the internal structure of an application component. A notable case is that of a distributed component featuring a parallel implementation in terms of multiple concurrent processes. In this situation structural changes consist in modifying the number of executed processes. This occurs when an insufficient number of nodes are available or if the task scheduler decides to change the set of processors allocated to a specific component.
- Implementation changes: Are modifications of the behavior of an application component, which changes completely its internal algorithm but preserving the elaboration semantics and the input and output interfaces with the other application components.
- Interface changes: Are intensive modifications of the component behavior, which completely changes the set of its provided external operations and services.

The existing programming models face the reconfiguration support development in two different ways. The first solution consists in putting everything concerning the adaptive behavior of a distributed application, like reconfiguration implementations, inside the run-time support system in such a way as to completely hide these aspects from the programmer standpoint. This transparent-approach requires a deep knowledge of the application structures so that the reconfiguration code can be automatically extracted by a static process of compilation of the program. In this scenario the programmer is freed from directly programming the reconfiguration phase but it is only involved in defining, by means of proper directives or programming constructs, the reconfigurations that are admissible (e.g. the compatibility between different versions of a component). On the other hand other programming models adopt a completely different approach to adaptivity, in which the programmer is directly involved in defining the reconfiguration activities. For each possible reconfiguration the programmer must provide the implementation that performs the dynamic reconfiguration, and it is also responsible for performing these activities in a fully correct and consistent manner.

Exploiting Adaptive Behaviors in Mobile and Pervasive Environments

Some research works provide programming frameworks for mobile applications that adapt their behavior in response to the actual level of resource availability (especially of communication networks). For instance in Odyssey mobile applications exploit run-time reconfigurations which are noticed by the final users as a change in the application execution quality. The Odyssey operating system is responsible for exploiting periodically resource monitoring activities and for interacting with mobile applications raising or lowering their corresponding quality level. In this approach all reconfiguration actions are automatically triggered by the run-time system without any direct user intervention. Nevertheless the most important drawback of this approach is the quite limited definition of the quality concept: In many cases it only consists in the quality of the visualized data but this assumption can be restrictive when we consider more complex applications involving an intensive cooperation between computation, communication and visualization phases.

In past researches, adaptivity consists in migrating specific parts of a distributed computation for optimizing performance and energy parameters as well as for reliability reasons. As an example in Aura adaptivity is expressed introducing the abstract concept of task: i.e. a specific work that a user has submitted to the system which can be completed by several distinct applications suitable for different classes of computing resources (e.g. workstations and smartphones). The framework is responsible for exploiting migration activities in a fully transparent way w.r.t the final users. Unfortunately this approach has been developed for very simple user tasks (e.g. writing a document or preparing a presentation). On the other hand, if we consider more complex mobile applications (e.g. that involve compute-intensive computations), transferring a partially computed task to a different platform can be a critical issue concerning both performance and consistency issues that have not been analyzed in these frameworks.

Recently the execution of high-performance parallel applications on mobile computing platforms have been addressed in some preliminary research works. A relevant example is MB++, a framework for developing compute-intensive programs for mobile and pervasive environments. In this approach one of the most important shortfalls is the quite limited utilization of mobile nodes, which are limited to pre-processing or post-processing activities whereas compute-intensive elaborations are executed only on HPC resources. In many time-critical scenarios, such as emergency management systems, we often require the possibility to execute forecasting models and decision support systems on a distributed set of localized mobile resources, equipped with a sufficient computational power.

Existing Programming Models for High-performance Adaptive Applications

Parallel computations are a particular class of tightly coupled distributed applications composed of several cooperating parallel modules. Adaptivity for parallel applications executed on traditional HPC architectures is an important feature, especially for real-time processing in which the presence of strong real-time deadlines and performance constraints require the capability of automatically adapting the application configuration for quickly responding to platform changes. Emergency management systems (e.g. disaster prediction and management, risk mitigation of floods and earthquakes) are notable examples in which the behavior of a time-critical processing needs to be periodically monitored and adapted throughout the execution.

One of the most widespread programming models for parallel applications is based on the MPI (Message Passing Interface) communication library. In this case parallel programs are expressed in an un-structured way by describing the behavior of a set of distributed processes cooperating by using communication channels. Last implementations of the MPI library provide some form of support to dynamic reconfigurations. MPI2 provides a mechanism for instantiating new processes at run-time: This feature can be exploited in such a way as to perform structural changes of a parallel program, for instance it is possible to increase the parallelism degree (e.g. the number of executed processes) achieving in this way an expected better performance. In general the management of adaptivity issues in MPI is completely left to the programmer, which is heavily involved in ensuring and maintaining the consistency and the correctness of the computation during and after the reconfiguration phase.

In order to partially solve the complexity issues of un-structured parallel programming, some notable research works have been proposed as the ASSIST programming environment. ASSIST is a general parallel programming framework for several classes of computing architectures, from shared-memory platforms as SMP and NUMA multi-processors to distributed-memory multicomputers as cluster of workstations and large-scale Grids. The most important novelty of this approach is the structured methodology for expressing parallel computations, which are instances of well-known parallelism schemes (e.g. task-parallel programs as task-farm or pipeline and data-parallel computations as map, reduce or communication stencils). This approach is known to the scientific community as Structured Parallel Programming.

In ASSIST a run-time support to dynamic reconfigurations is rendered by exploiting a transparent-approach to adaptivity. A dynamic reconfiguration is a change involving a specific application component by modifying: (i) The mapping between execution processes and underlying computing resources (i.e. geometrical changes); (ii) by increasing or decreasing the number of execution processes of a component (i.e. structural changes). Some papers describe how adaptivity is exploited in the ASSIST framework. Reconfigurations are performance-oriented: dynamic changes of application components are triggered by the run-time support in presence of QoS violations of a pre-defined performance contract.

Another interesting work is the Behavioural Skeleton approach. Adaptivity for distributed high-performance computations is exploited by means of the Grid Component Model (GCM) and the structured parallel programming paradigm (which the authors also call skeleton-based programming). In GCM a self-adaptive component is composed of two main parts: the Membrane which is an abstract unit responsible for controlling the adaptive behavior of the component, and the Content composed of a set of processes performing the corresponding functional logic of the computation. These entities can also be other GCM components (i.e. inner components): Therefore the GCM model makes it possible a natural hierarchical nesting between several self-adaptive components.

This approach is characterized by very interesting run-time mechanisms for controlling multiple non-functional concerns (e.g. it is possible to simultaneously control different parameters as performance and security objectives). In this case the solution proposed in provides multiple autonomic managers for a single component, each one controlling a specific non-functional concern by using a set of event-condition-action (ECA) policy rules. Different policies can lead to some conflicting decisions: in this case the authors propose distributed consensus-based solutions.

Application Scenarios

We are interested in describing the design and the developing issues of time-critical systems as emergency management applications. They are a notable example of systems that exploit highly heterogeneous distributed computing platforms, composed of traditional high-performance architectures as well as mobile nodes and sensors, providing system features and services according to precise QoS constraints. Such QoS is typically related to the performance at which computing results or communication facilities are provided to users and/or to the service availability and reliability w.r.t. software and hardware failures.

A main issue in executing this class of applications on heterogeneous platforms is hence given by the chance of defining proper programming models and run-time supports aiming at enabling the definition and dynamic satisfaction of QoS constraints. This issue can be seen as even more complex by considering the strong dynamic variability of computing and communication services provided by these platforms, also given by the mobility of some of its nodes and their geographical distribution.

Emergency management systems include data- and compute-intensive processing e.g. forecasting and decision support models) not only for off-line centralized activities, but also for on-line and decentralized activities. Consider the execution of software components performing a forecasting model, which is a critical compute-intensive computation to be executed respecting precise operational real-time deadlines. In "normal" conditions (e.g. concerning the network availability and the emergency scenario itself), we could be able to execute these components on a centralized server, exploiting its processing power to achieve the highest performance possible. Critical conditions in the application scenario (e.g. an emergency detection) can lead to different user requirements (e.g. increasing the performance to complete the forecasting computation within a given, new deadline). Moreover, changes in network conditions (e.g. of the interconnection network with HPC centralized resources) or in computing resource availability can lead to the necessity to execute proper versions of the application components directly on localized mobile resources which are available to the users (e.g. civil protection personnel, rescuers and stakeholders).

In such cases, compute-intensive elaborations that need to be periodically executed for monitoring, predicting and taking response actions during an emergency phase, can be alternatively executed on different or additional computing resources, including sets of distributed mobile resources running properly developed versions of these computations. In other words, in these scenarios it is important to assure the service continuity, adapting the application to different user requirements but also to the so-called execution context, which corresponds to the actual conditions of the both the surrounding environment and the computing and communication platform. So the key issues in the definition of high-performance parallel programming paradigms, models, and frameworks to design and develop these kinds of complex and dynamic applications, is adaptivity (possibility to adapt the application behavior changing component versions and the platforms on which they are executed) and Context Awareness (knowledge about the current condition of the reference environment).

The general reference point for these kinds of applications is the Grid paradigm which, by definition, aims to enabling the access, selection and aggregation of a variety of distributed and heterogeneous resources and services. However, though notable advancements have been achieved in recent years, current Grid technology is not yet able to supply the needed software tools to match the high-performance feature with high adaptivity, ubiquity, proactivity, self-organization, scalability, interoperability, as well as fault tolerance and security, of the emerging applications running on a very large number of fixed and mobile nodes connected by various kinds of networks.

We claim that a high-level programming model is the only viable solution to design and develop such kind of distributed applications. A novel approach needs to provide a proper combination of high-performance programming models and pervasive and mobile computing frameworks, in such a way to express a QoS-driven adaptive behavior for critical high-performance applications on heterogeneous contexts.

Features and Requirements

A novel approach needs to be characterized by a strong synergy between two different research fields: Pervasive and Mobile Computing and Grid Computing. Pervasive and Mobile Computing is centered upon the creation of systems characterized by a multitude of different computing and communication resources, whose integration aims at offering seamless services to the users according to their current and time-varying needs and intentions. On the other hand Grid Computing focuses on the efficient execution of compute-intensive processes by using geographically distributed sets of computing resources.

To merge these two areas, an effective integration must provide a proper combination of high-performance programming models and mobile computing frameworks in such a way as to express a QoS-driven adaptive behavior for distributed applications like emergency management systems. In this section we will highlight the main guidelines that need to be followed for exploiting such integration. They can be summarized in three different points: (i) How compute-intensive processing for heterogeneous environments can be programmed and their performance formally analyzed; (ii) how a distributed application can change its QoS behavior at run-time; (iii) when it is necessary to execute an application reconfiguration.

Structured Parallel Programming

First of all we require a structured methodology for expressing parallel and distributed programs in such a way as to make the programming effort less costly and less time-consuming and to have a predictable QoS behavior for such applications. This means that we need sufficient high-level abstractions featuring a large-degree of programmability, compositionality and performance portability. Portability plays a central role in heterogeneous contexts: portable parallel applications are able to achieve acceptable performance results on different computing platforms, trying to exploit in the best way as possible the underlying physical aspects of the target machine. This can be done by tuning proper application parameters (e.g. parallelism degree and task granularity) statically, during the compiling phase, or dynamically through run-time reconfiguration processes.

For these reasons Structured Parallel Programming (SPP) has been proposed as an effective and attractive approach to parallel programming featuring interesting properties in terms of high-level programmability and performance portability. According to the SPP methodology a parallel computation is expressed by using well-known abstract parallelism schemes for which parametric implementations of communication and computation patterns are known. In fact the experience in parallel programming suggests that parallel programs make use of a limited number of parallelism patterns exhibiting regular structures, both concerning data organization and partitioning or replication of functions. In this way we can identify several parallelism paradigms as data parallelism schemes (e.g. map, reduce, parallel prefix and communication stencils) and task parallelism structures (e.g. pipeline, task-farm and data-flow). Furthermore the performance behavior of these parallelism schemes has been studied by exploiting formal analysis (i.e. performance models) based on queueing theory and queueing network fundamental results, thus rendering the performance modeling of this class of computations usable also from automatic tools as compilers (e.g. for statically deciding the best application configuration) and from run-time supports (e.g. for providing efficient fault-tolerance mechanisms but also for deciding the execution of dynamic reconfigurations.

The SPP methodology is based on the concept of parallelism schemes that exhibit the following features: (i) They are characterized by constrains in the parallel computation structure; (ii) they have a precise semantics; (iii) they are characterized by a specific performance model; (iv) they can be composed each other to form complex graph computations. A central property of these schemes is the complete knowledge of their computation structure, e.g. in terms of data and functions replication and/or partitioning. For instance task-farm paradigm exploits replication only, applied to functions and to non-modifiable data. Pipeline exploits a partition of the sequential elaboration in a sequence of multiple successive phases each one using a set of replicated (non-modifiable) or partitioned (modifiable) data. On the opposite direction data-parallel schemes correspond to the replication of the same functionality and to the partitioning of data, so that distinct parallel units are able to apply the same operations to different data partitions in parallel.

Performance of parallel computations can be measured in terms of throughput, i.e. completed tasks per time unit, and computation latency, i.e. the average time needed to execute the computation on just one input task. Parallelism schemes can have different impacts on these two performance measurements. Task-farm and pipeline, though able to increase the throughput of the computation, also tend to increase the latency compared to the sequential implementation due to communication overhead. Data-parallel and data-flow tend to improve both the computation latency and the throughput, but they can be hampered by load-balancing issues. Finally parallelism schemes are also characterized by a different degree of memory utilization. Task-farm tends to increase the memory capacity, the others, being based on data partitioning, tend to decrease it.

As far as composability is concerned, parallel distributed applications can be represented as graphs of modules. Each module can exploit a sequential or a parallel computation based on a specific parallelism scheme (intra-module parallelism), whereas modules can be composed in general complex graph structures (inter-module parallelism).

Dynamic Reconfigurations of Parallel Programs

Previously in this chapter we have introduced the concept of application configuration. Any run-time activity that modifies the parallelism degree, the parallel version (e.g. parallelism scheme and parallelized sequential algorithm) or the execution platform of a module is indicated as a dynamic reconfiguration process.

Although there are specific situations in which the "best" configuration for each module can be statically selected during the system design phase (e.g optimizing performance parameters and memory and energy usage), in dynamic execution conditions this best configuration might not be identifiable without run-time information. As an example consider the following situations:

- In dynamic environments, the degree of availability of computing and network resources can dynamically vary also in unexpected ways.
- For compute-intensive elaborations (as forecasting models and simulations), precise user requirements can be requested. For instance when an emergency scenario is detected, the decision support model that assists civil protection personnel could require to be executed until a maximum completion time.
- For irregular parallel problems, characterized by computations whose size, distribution and complexity depends on the properties of input data, a run-time support to dynamic reconfigurations is an unavoidable feature that needs to be provided.

For structured parallel computations we can classify the set of adaptation processes involving a single application module in two categories namely non-functional and functional reconfigurations.

Non-functional Reconfigurations are adaptation processes involving the run-time modification of some implementation aspects of a parallel module of the application graph. For implementation aspects we intend that:

- It is possible to modify the current parallelism degree of a module, e.g. increasing the number of implementation processes in such a way as to achieve a better performance level (e.g. response time for each request or the completion time of the entire computation).
- The run-time support can modify the mapping (i.e. geometrical changes) between implementation processes of a parallel module and underlying physical computing resources on which they are executed.
- The run-time support can provide another important class of geometrical changes involving the run-time data re-distribution among the set of worker processes

of a parallel module. Such changes are an effective approach for addressing potential workload un-balancing problems of data-parallel parallelizations.

The most important aspect characterizing non-functional reconfigurations is that in every case these processes do not modify the sequential algorithm performed by the parallel module, neither the exploited parallelization scheme.

Another important and, in some sense, intuitive concept is that the same problem can usually be solved in a parallel fashion by exploiting several different parallelization schemes. For instance let us consider a simple example in which a parallel module periodically receives a sequence of input tasks each one consisting in a pair of a matrix and a vector. For each task the module performs a matrix vector product and the result vector is transmitted to a destination module. According to the way in which we perform data distribution among parallel entities, we can parallelize this problem in several ways. A task-farm scheme consists in replicating the entire matrix-vector computation on multiple independent workers that apply the computation on a different set of scheduled input tasks. In this way we are able to improve the throughput of the computation but not the computation latency. For this simple example several data-parallel schemes can also be adopted. For each input task, if we partition the input matrix among a set of parallel processes and we replicate the input vector, we are able to exploit a classic data-parallel map scheme in which each process exploits the matrix-vector computation on its partition without requiring inter-process data exchange and synchronizations. Moreover, if the input vector instead of being replicated is partitioned among processes, we are able to exploit a data-parallel approach with a communication stencil. In both the cases the data-parallel approach improve latency of the computation as well as its memory occupation w.r.t the task-farm implementation. Moreover the stencil version requires a higher number of exchanged data during the computation.

The previous scenario suggests that we can use this important feature of structured parallel programming in such a way as to effectively deal with highly dynamic and heterogeneous execution environments. For each application module several alternative versions can be provided, and Functional Reconfigurations are implementation changes that consist in changing at run-time the version which is currently executed by a parallel module. Versions can have a different but compatible semantics: They can exploit different sequential algorithms, parallelization schemes or optimizations, but preserving the module interfaces in such a way that the selection of a specific version does not modify the global application behavior. Each version of the same parallel module is optimized for being executed whenever certain execution conditions are satisfied (e.g. depending on the actual levels of communication bandwidth, power supply, available memory) or on the presence of specific classes of computing resources.

Another important advantage of this methodology w.r.t non-structured parallel programming models is that it renders feasible a transparent approach to application reconfigurations. The reconfiguration code that performs non-functional reconfigurations (as parallelism degree variations) can be automatically extracted from the knowledge of the parallelism scheme that needs to be modified. In this case the programmer is only involved in providing the implementation of the high-level functions executed by a set of worker processes, and their number and the interconnections for task reception and result transmission are completely managed by the run-time support system. Moreover, for functional reconfigurations, in formal consistent reconfiguration protocols have been presented for switching between different structured parallelism schemes.

Exploiting Adaptation Strategies for Parallel Programs

An adaptive application is a system which evolves over time changing its initial configuration in response to external events. Providing a run-time support to dynamic reconfiguration activities is a first and essential requirement for developing adaptive high-performance applications. Although many research works focus on specific implementation issues describing several optimizations for reducing reconfiguration costs in terms of performance degradations for certain parallelization schemes and their compositions, they do not pay sufficient attention to the decision process that triggers the execution of these reconfigurations. We refer to this process as an Adaptation Strategy, that is a plan or a set of rules used by the system for achieving the specific objectives of its adaptive behavior.

Especially for emergency management systems several constraints need to be satisfied throughout the execution. The system must offer its functionalities to the users according to different QoS notions:

- In many adaptive systems we need to control the execution progress preserving observed metrics within user-defined ranges. Classic examples are: e.g. maintaining the mean response time of the system within a window of tolerance, i.e. between a maximum and a minimum acceptable threshold over the execution. In this case we refer to this objective as a threshold specification problem.
- Alternatively we might require to maintain some quantitative execution metrics as closer as possible to a set of desired reference values. For instance it could be necessary to maintain the mean service time of a computation equal to a specified target value desired by the final user. We refer to this approach as a set-point regulation problem.
- More often we need to control several execution measurements of the system, as the performance, energy and memory requirements, number of utilized computing and network resources. In this situation we need to find a control law such that a certain optimality criterion is achieved. This problem can be casted into a mathematical fashion introducing an objective function $F(x_1, x_2, \ldots, x_n)$ of n input parameters whose value should be minimized or maximized. Inputs are

system controllable features, and the adaptation strategy is aimed at optimizing this multi-variable function during the system execution. This formulation is known as an utility/cost optimization problem.

In different research fields (e.g. Autonomic Computing and Control Theory) general adaptive systems have been described in terms of a reference control architecture composed of two main parts and proper interaction phases between them. The two parts are:

- The target controlled system (namely plant) that we want to control. It takes functional inputs and generates functional outputs according to the semantics of its computation. However, for controlling the behavior of a system, other relationships among non-functional inputs and outputs are also important. System execution can be influenced by a specific set of manipulated control inputs, whose values strictly affect the way in which the system execution is exploited. Based on the desired control objectives, the quality of system operation is evaluated by measuring non-functional outputs also called observed outputs (e.g. the current level of some performance metrics).
- A controller, which is an independent entity able to monitor and affect the system operational conditions by taking and analyzing observed outputs and promptly decide the corresponding set of control inputs.

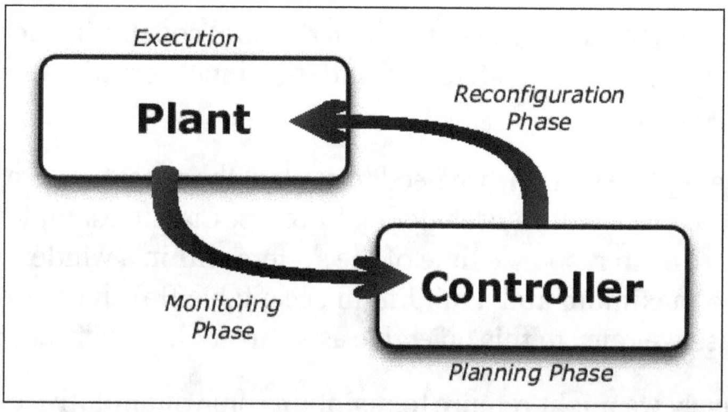

Control-loop scheme of an Adaptive system: The cycle evolves according to a monitoring phase, a planning phase and a reconfiguration phase.

Interactions between these entities follows a classic closed-loop feedback scheme composed of different phases. Monitoring phase involves capturing current properties and measurements of the system which are effective for identifying when the execution of a dynamic reconfiguration can be useful for achieving the desired objectives. Planning phase consists in a set of concrete actions aimed to select a new set of control inputs that are the best response to the current observed outputs of the system. Finally the Reconfiguration phase applies the decided set of control inputs to the controlled system. In this part of the chapter we are interested in understanding how the planning phase can be exploited and what methodologies can be used.

Adaptation Strategies Expressed by Logic Rules

A flexible methodology for expressing adaptation strategies, which has been exploited in Pervasive and Mobile scenarios due to its intrinsic simplicity and programmability, consists in defining adaptation processes as reactions to specific system situations. As an example if the available energy level of a mobile device is lower than a specified threshold (e.g. 25 %), the execution of a mobile application can switch to a low-energy version which preserves the battery duration employing a limited utilization of the device display. Therefore a straightforward solution for expressing adaptation strategies consists in providing a mapping between events and execution situations and corresponding reconfigurations (i.e. situation-action pairs) as a finite set of imperative logic rules. System administrators and designers encode management guidelines as a set of rules that represent the adaptation policy of the system.

Policy rules are a form of guidance used to determine decisions and corresponding actions on the system execution. They have been introduced especially in the field of intelligent agents Russell & Norvig, where abstract entities are able to perceive their environmental conditions and act on the basis of this information in order to maximize their objectives. In this vision an adaptive system is in a specific internal state at every given moment of time, and a set of logic rules may cause an action to be taken an therefore a transition to a different internal state of the system.

Logic rules can be expressed according to several paradigms. Although many techniques exist, with specific modifications and features depending on the particular application context, a general approach is based on Event-Condition-Action (ECA) policy rules for programming reactive behaviors. ECA rules have the following basic syntax: when event if condition then action. Informally the abstract semantics is: the occurrence of the event triggers the rule evaluation, the condition is checked in order to ensure that the system is in a specific internal state. If this condition holds, the corresponding action is enforced. Notice that though the state that will be reached by applying the rule is not explicitly expressed, the policy programmer knows the desired effect of the selected action.

If multiple rules are fired simultaneously, the entire set of corresponding actions will be performed at the same time. This may lead to potential conflicting actions (e.g. resulting in conflicting post-adaptation situations). The design of proper techniques for efficiently identifying and resolve such conflicts has been an intensively studied research issue. Although the large research effort, the general problem of avoiding conflicting rules is hard to be solved, especially for complex systems in which the consequence and the side-effects of a dynamic reconfiguration involve the long-term steady-state behavior of the system (i.e. in this case a rule can conflict with rules that may be triggered on different events in future). This suggests that, although highly programmable and user-oriented, adaptation strategies based on logic rules are difficult to be applied to complex applications as emergency management systems. In fact in this case it is extremely important to have a reasonable expectation about the consequences of reconfiguration actions on the long-term behavior of the systems (e.g. future performance level after the completion of a reconfiguration process), that can be difficult to be evaluated and represented through situation-action pairs.

Adaptation Strategies based on Control-theoretic Techniques

Besides Artificial Intelligence and Reactive Systems, the concept of automated operations is an important research field in disciplines as mechanical and electrical engineering for developing autonomous systems able to respond to changing workloads and expected run-time conditions. The methodology based on Control Theory foundations has been intensively exploited for design controllers and feedback systems in many industrial plants and mechanical infrastructures. These solutions provide powerful mechanisms for dealing with unpredictable changes, uncertainties, and system disturbances.

In many research works several control-theoretic modelings of computing and networking systems have been proposed. Control-based approaches are often a solid solution to solve network problems like congestion and flow control, and rate adaptation of queueing networks. This leads to an increasingly important research area in which the adaptive system is the computing network itself, i.e. Autonomic Networking, whose interconnection facilities are able to automatically detect, diagnose and repair failures, as well as to adapt the underlying network configuration and optimize its performance and quality of service.

Nevertheless the exploitation of control theory methodologies to computing systems is rarely used in practise. The formal design of controllers needs the ability to precisely quantify the reconfiguration effects on the system evolution. Hence, system modeling is a prerequisite for applying such techniques. This model needs to be expressed in an input-output form concerning how control inputs and disturbances affect the behavior of observed measurements. Such relationships between model variables can be determined through first-principle models, based on the actual physics laws which govern the evolution of the system, or exploiting empirical models in which the relationships are extracted through static techniques (e.g. least-square techniques) on properly defined experimental data and observations.

We claim that a relevant example in which first-principle models can be used consists in the performance modeling of structured parallel computations. For these structured parallelization schemes and their composition in computation graphs, performance models provide a quite acceptable performance predictions of system steady-state behavior, which is often a sufficient guideline to define powerful control-theoretic strategies for controlling parallel applications. When it is possible, these approaches are amenable to provide properties like the accuracy of an adaptation strategy (how precisely control objectives are satisfied), settling time (how long a reconfiguration lasts until the steady-state behavior of the system is reached) and the stability degree of a reconfiguration (how long a system configuration represent a "good choice" for achieving the desired system objectives).

For these reasons control theory techniques, though their applicability is not always straightforward, represent a valuable research direction for controlling time-critical applications as emergency management systems. Based on the main guidelines described in this section, in the rest of this chapter we will introduce in more detail our approach.

Design Principles of a Novel Programming Model Approach

Whereas the structure of the graph can be arbitrary (e.g. cyclic client-server interactions or acyclic graphs of multiple computation phases), the internal parallelism inside each module is expressed by means of structured parallelism schemes.

The adaptive behavior is expressed in the possibility of each parallel module to perform dynamic reconfiguration activities triggered by their own control logic. This means that adaptivity is exploited through a set of independent entities that cooperate for implementing the distributed functional logic of the application and also for negotiating reconfigurations taken by their control logics.

Structure and Modeling of an Adaptive Parallel Module

The core element of our approach is the concept of adaptive parallel module (shortly ParMod), an independent and active unit featuring a parallel elaboration and an adaptation strategy for responding to different sources of dynamicity. From an abstract point of view a ParMod can be structured in two interconnected parts:

- An Operating Part is responsible for performing a parallel computation expressed according to a certain parallelism scheme of the SPP framework (e.g. task-farm and data-parallel are relevant examples). Without loss of generality we assume a stream-based computation in which a set of input data streams from other parallel modules are received by the operating part. The parallel elaboration is activated according to a non-deterministic selection of input elements or, based on a data-flow semantics, waiting for the reception of an element from each input interfaces of the module. Result externalization (if it is necessary) is exploited onto output data streams (output interfaces) to other parallel modules.
- A Control Part is an autonomous entity able to observe the operating part execution and modify its behavior exploiting reconfiguration activities.

For each ParMod we suppose the presence of multiple alternative configurations of its operating part that differ in the parallelism degree, the structured parallelism scheme or in the execution platform on which they can be deployed. The only constraint that we impose is that every configuration must respect the same input and output interfaces of the ParMod, in such a way that a local reconfiguration involving a single parallel module does not modify the interfaces provided to other application modules. Therefore each ParMod features a multi-modal behavior defined as follows:

At each point of time the operating part can behave according to a certain active configuration belonging to a specific set C of alternatives:

$$C = \{C_0, C_1, \ldots, C_{v-1}\}$$

This set represents a finite and discrete set of statically known alternative configurations of the parallel module.

The adaptive behavior of a parallel module is exploited through a periodical information exchange between operating and control part. Observed outputs are QoS measurements (e.g. memory occupation, energy consumption, number of completed tasks) that are periodically monitored by the control part. Control inputs are proper commands that modify the current operating part configuration exploiting non-functional or functional reconfiguration activities. Observed outputs and control inputs exchange can be exploited periodically, at fixed points equally spaced (in this case we speak about a time-driven controller) or whenever specified events are satisfied (as in the case of event-driven controllers).

Our ParMod control model is based on a time-driven control part that takes reconfiguration decisions every control step of duration τ. In other words at the beginning of each control step the control part acquires observed outputs from operating part, executes a specific control algorithm to decide a corresponding set of control inputs that will be communicated to the operating part.

A Hybrid Modeling of the Operating Part Behavior

In order to apply a formal methodology for developing ParMod adaptation strategies, we need to model the QoS temporal evolution of a parallel computation by using a mathematical and formal approach. This means that we need proper mathematical equations that express the relationship among observed output measurements and current control inputs. The structured parallel programming paradigm is a basic point in order to establish such modeling. Relevant examples are:

- Structured parallelism schemes feature a sort of predictability of their steady-state performance level. Fixed the parallelism degree and the execution platform, several performance measurements as the mean service time, the mean queue length and the computation latency can be predicted and formally analyzed according to analytical models based on queueing theory and queueing network results.

- Memory utilization models can also be derived for well-known parallelism schemes by exploiting the specific behavior of different parallelization patterns in terms of function and data replication or partitioning.
- Structured parallelism schemes have a precise semantics in terms of computations performed by each parallel unit, the size of exchanged messages and the frequency of certain activities as calculation and message transmission. Such knowledge can be a starting point in order to define models for measuring the power consumption of a parallel module, especially when its execution is mapped onto energy-limited resources as mobile devices.

A formal model that describes the predictability of QoS parameters involves the following set of variables:

- State variables represent system measurements that are useful to maintain during successive control steps. These variables can describe the global energy consumption caused by the parallel execution of a ParMod, the number of queued tasks, the global number of completed input stream elements. We refer with the term x(k) as the current value of the internal state of the operating part model at the beginning of the control step k. State variables are directly monitored at each control step by the control part.
- Disturbance variables are uncontrolled exogenous signals that can affect the relationship between control inputs and the observed state variables modeling the actual QoS of the computation. Uncontrolled means that the control part is not able to decide and fix their values, but they are determined by environmental or external decisions outside of ParMod control. Relevant examples are the mean calculation time of a sequential function on a certain target architecture, the communication latency and the mean inter-arrival time of requests to the system. We refer with the term d(k) as the value assumed by disturbances throughout the k-th control step.
- Control inputs, u(k), indicate the ParMod configuration that will be exploited throughout the k-th control step.

According to the multi-modal behavior of the operating part and the QoS predictability of different ParMod configurations, we can identify two classes of transitions that characterize the ParMod execution:

- Continuous transitions: When a configuration has been fixed, the evolution of continuous-valued QoS state parameters can be predicted by applying a specific mathematical model corresponding to the currently used configuration.
- Discrete transitions: By executing reconfiguration activities, the current configuration can be changed passing from a configuration C_i to a different alternative configuration C_j, shortly $C_i \rightarrow C_j$ with $i \neq j$.

The presence of continuous transitions (of continuous variables) and discrete transitions (of alternative configurations) suggests to model the operating part of a ParMod as an hybrid system in which these two dynamics are formally modeled in a unique and refined mathematical structure.

For each configuration Ci ∈ C the temporal evolution of observed outputs and of the next state variables can be expressed by a model in a state-space form:

$$\mathbf{x}(k + 1) = \varphi_i\ \mathbf{x}(k), \mathbf{d}(k)$$

The semantics of the next state expression provides that, if the values of expected measured disturbances and the current model state for control step k are known, by applying a function φi we can predict the values assumed by state variables at the beginning of the next control step k + 1.

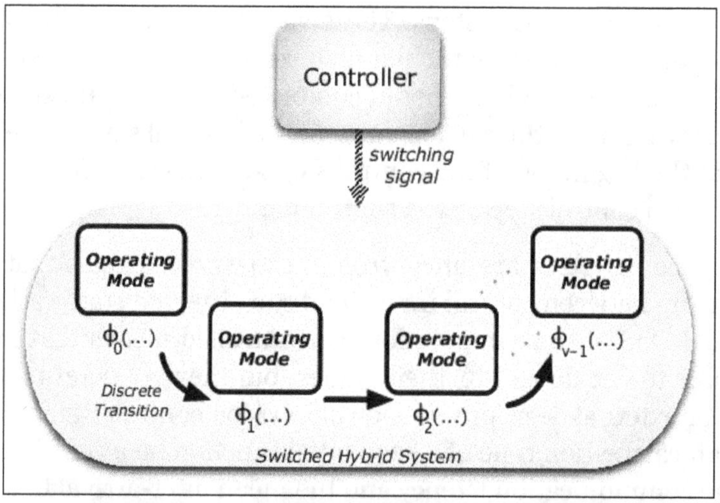

The operating part evolves according to continuous transitions (i.e. concerning QoS variables predicted through the model of the current configuration), and discrete transitions.

Two considerations are important at this point. First of all the function φi can be obtained by applying statistical methods or, as we have seen, by using a higher level knowledge of the structure of the computation behavior in order to extract first-principle relations. The second point is that each model φi is strictly coupled with a configuration of the ParMod. In some cases a unique mathematical model can be parametrically provided for multiple configurations (e.g. as for different parallelism degrees), but there are also situations in which configurations can require a completely different modeling of the state variables dynamics (e.g. as for the case of parallel versions expressed by means of different parallelism schemes as task-farm and data-parallel ones).

This variable-structure behavior of the operating part is typical of a large class of hybrid systems featuring a limited set of alternative operating modes (i.e. configurations in our case). Such class consists in the so-called Switched Hybrid Systems Liberzon in which control inputs from the controller modify the actual configuration of the operating part, and future QoS measurements can be predicted following the evolution rule strictly coupled with the new configuration. In conclusion the entire operating part model is given by:

$$\mathbf{x}(k+1) = \Phi(\mathbf{x}(k), \mathbf{d}(k), \mathbf{u}(k))$$
$$= \text{if } (\pi\, \mathbf{u}(k) = Ci) \text{then } \varphi_i\ x(k), \mathbf{d}(k)$$

in which function Φ is a piecewise-defined function that provides a different definition for each alternative configuration of the parallel module, whereas π is a bijective function that maps each possible value of control inputs onto corresponding configuration indices.

A Control-theoretic Adaptation Strategy for ParMods

In this section we will propose the exploitation of a control-theoretic technique for controlling the QoS behavior of an adaptive parallel module. We point out that in the context of time-critical applications, like emergency management systems, a classic reactive approach in which the system reacts to well-identified circumstances in pre-programmed way (as in classic environments for adaptive HPC applications) is not an effective adaptation strategy. The criticity of QoS requirements for these applications imposes that reconfiguration decisions must be selected taking into account the history of the system, trying to respond in a pro-active fashion to the user requirements and to the future behavior of the surrounding execution environments. In this case we need adaptation strategies that are more similar to the ones studied in pervasive systems, in which taking actions in advance to potentially critical future events is a central point of this approach.

The predictive control approach that we propose for adaptive parallel applications is based on the following points:

- The presence of (empirical or first-principle) mathematical models that can be used to estimate future QoS behavior of the computation in function of a planned sequence of control inputs (i.e. a reconfiguration plane) and future predictions of disturbances.
- The presence of an objective function describing the control aims which drive the selection of the optimal control input sequence.
- The specification of boundary conditions on state and control input variables.

This formulation of the predictive control problem is known to control theorists as Optimal Control. Optimal control is the process of determining control and state trajectories for a system over a certain period of time in order to optimize a properly defined objective function.

A typical representation of an objective function is given below:

$$\max U(k) = \sum_{i=k}^{k+h-1} L(\mathbf{x}(i+1), \mathbf{u}(i))$$

The function represents an aggregate utility (or cost) which depends on the desirability of future system internal states and control inputs taken for a horizon of N successive control steps (e.g. ideally the whole execution duration). Expressing control preferences among different future states and control input trajectories is an expressive way to define powerful control strategies for structured parallel computations featuring a predictable behavior. Objective functions can be tuned to satisfy the specifications of the system, the control objectives as well as to find feasible trade-offs between conflicting requirements (e.g. optimization problems can exploit an intrinsic multi-variable nature including performance, power and memory measurements and specific constraints on state and control variables).

If we suppose to know the exact trajectory assumed by disturbance inputs for the whole execution duration, and if the system model is sufficiently accurate and precise, we can statically define an optimal reconfiguration plane that optimizes the ParMod execution. Of course this assumption is not always feasible for the following reasons:

- Disturbance inputs are variables whose behavior (e.g. average values) can not be statically known a priori but they may depend on uncontrollable factors (as the actual conditions of the underlying execution platform).
- The system model can be effected by perturbations and unmodeled dynamics (e.g. due to unmeasured disturbances) that limit the quality of the future QoS estimations.

For the previous reasons applying the optimal reconfiguration plane step-by-step in an open-loop fashion is not a viable approach. Nevertheless several sub-optimal approaches are available in order to iteratively apply optimization problem solutions in a closed-loop fashion.

Model-based Predictive Control of Structured Parallel Computations

Model-based predictive control (shortly MPC) is a repetitive procedure that combines the advantages of long-term planning (feedforward control based on system predictions over a future horizon) with the advantages of a feedback control using actual system and disturbance measurements. At the beginning of each control step k the values of the model state variables and of past disturbances are measured by the control part. At this point a limited future time horizon (i.e. a prediction horizon)of h consecutive control steps is considered, and a prediction of disturbances for this short time interval is exploited through proper statistical techniques.

The predicted disturbance trajectory is composed by:

$$D_k^{k+h-1} = \left\{ \hat{\mathbf{d}}(k \mid k), \hat{\mathbf{d}}(k+1 \mid k), \ldots, \hat{\mathbf{d}}(k+h-1 \mid k) \right\}$$

where the syntax $\hat{\mathbf{d}}(k+1 \mid k)$ means that the estimated value of disturbance inputs for the step k + i is predicted using the current knowledge at control step k.

These predictions are used to plan a sequence of optimal reconfiguration decisions for each control step of the prediction horizon. The optimization problem is solved for the limited time horizon finding an optimal control input trajectory:

$$U_k^{k+h-1} = \left\{ \mathbf{u}(k \mid k), \mathbf{u}(k+1 \mid k), \ldots, \mathbf{u}(k+h-1 \mid k) \right\}$$

Instead of applying the optimal reconfiguration plane step-by-step, the uncertainty of disturbance estimation (which increases going deeper in the prediction horizon) and the potential inaccuracy of the future QoS predictions, suggest a more effective approach based on an iterative procedure. Only the first control decision of the optimal trajectory is transmitted from the control part to the operating part at the beginning of the current step k while the rest are discarded. This process is repeated at the beginning of the next control step when: (i) the new system state is available; (ii) values of past disturbances for the previous step have been measured. The effect of this control algorithm is to move the prediction horizon towards the future following the so-called receding or rolling horizon technique.

Based on the QoS predictability of structured parallelism schemes, the benefits of the predictive approach are evident. Suppose to have a ParMod control objective formulated as a threshold specification problem: e.g. we need to maintain the observed mean throughput (completed tasks per time unit) of a parallel computation within an acceptable region of values (e.g. between a maximum and a minimum threshold established by the user). Although an initial configuration (e.g. parallelism degree) can be identified during the design phase such that this QoS requirement is met, future modifications of the initial conditions modeled as disturbances (e.g. time-varying task grain) can prevent this configuration to be no longer acceptable. Especially for time-critical applications the utilization of a predictive strategy can be a valuable solution for two main reasons:

- Selecting in advance reconfiguration actions can be crucial in order to anticipate undesired behaviors and promptly mitigate the future variability of disturbances.
- Typically reconfigurations involve a cost on the execution. After a reconfiguration a ParMod can incur some dead-time, e.g. for completing reconfiguration protocols or rather the cost for changing the number of currently used computational resources (e.g. as in Cloud Computing environments).

Despite MPC is a largely used approach in many real-world scenarios (e.g. for controlling chemical, mechanical and industrial plants), it is considered a compute-intensive technique especially for hybrid systems (as a ParMod) where the set of control inputs is discrete. In this case the on-line optimization problem requires to explore a search space whose size grows exponentially with the length of the prediction horizon. The search space can be represented as a complete tree (i.e. the ParMod evolution tree) with a depth equal to the prediction horizon length h and an arity that coincides with the number of ParMod configurations v. The number of explored nodes is given by $\sum_{i=0}^{h} v^i$ thus exponential in the prediction horizon length. If not properly addressed with specific techniques (e.g. branch&bound approaches) or heuristics, the combinatorial optimization problem that needs to be solved at each control step implies an exhaustive search by testing among all the feasible combinations of reconfigurations, thus potentially limiting its viability to systems with long sampling intervals. Nevertheless in practical scenarios, since the number of ParMod configurations is often sufficiently limited and prediction horizons are normally short due to disturbance prediction errors, this approach can be exploited in practise without complex search space reduction techniques.

The application graph is depicted in figure. The system is responsible for providing short-term forecasting results of a flood scenario in a real-time fashion to a set of interested users (e.g. emergency stakeholders or rescuers). The graph is composed of a Generator module that periodically produces sensor data (e.g. water surface elevation and water speed) for each point of a 2D discretization of the emergency scenario (e.g. a river basin). Each point is considered as an independent task by an adaptive Solver ParMod, that numerically solves a system of differential equations describing the flow behavior at the surface. Results evaluation is exploited by a Client component that performs post-processing and visualization activities on the mobile devices of the users.

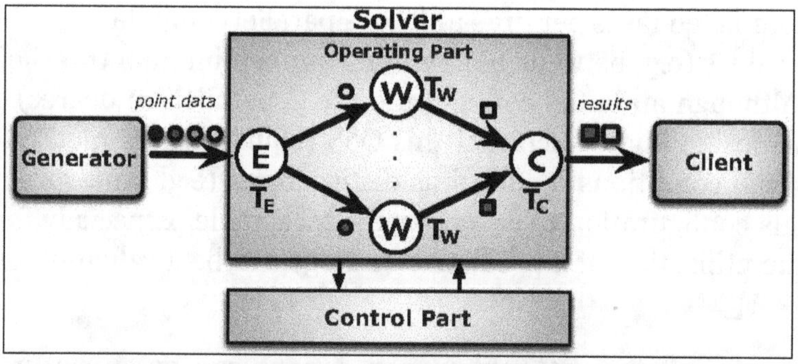

Flood Emergency Management Application: This test-bed scenario considers a limited set of application modules.

An example of flood forecasting model is the TUFLOW hydrodynamic model Charteris et al., which is based on mass and momentum partial differential equations to describe the flow variation at the surface. Their discrete resolution requires, for each time slice, the resolution of a very large number of sparse tri-diagonal linear systems. For their resolution several highly scalable techniques have been developed over the last decades. In we have described two sequential variants of the cyclic reduction algorithm.

We have provided two distinct parallel versions of the Solver ParMod: The first variant of the cyclic reduction, which requires a smaller number of operations than the second one, is executed inside a task-farm scheme, in which an emitter entity schedules input tasks (points) among a set of workers that execute in parallel the computation on different stream elements. The second version consists in a data-parallel version of the second variant. Although this algorithm requires a greater number of operations, data dependencies and also a higher memory utilization, it is amenable to be implemented in parallel through data-parallel techniques, in which a set of workers cooperatively compute the calculation on a partition of the same input data-structure. Moreover, in order to implement the data dependencies imposed by algorithm semantics, workers communicate with each other following a statically known communication stencil whose form varies between different iterations of the computation.

Such parallelizations respond to different QoS requirements: Task-farm is able to increase the throughput of the computation in terms of number of completed points per time unit, whereas, the data-parallel scheme, can also improve the computation latency for completing a single task elaboration.

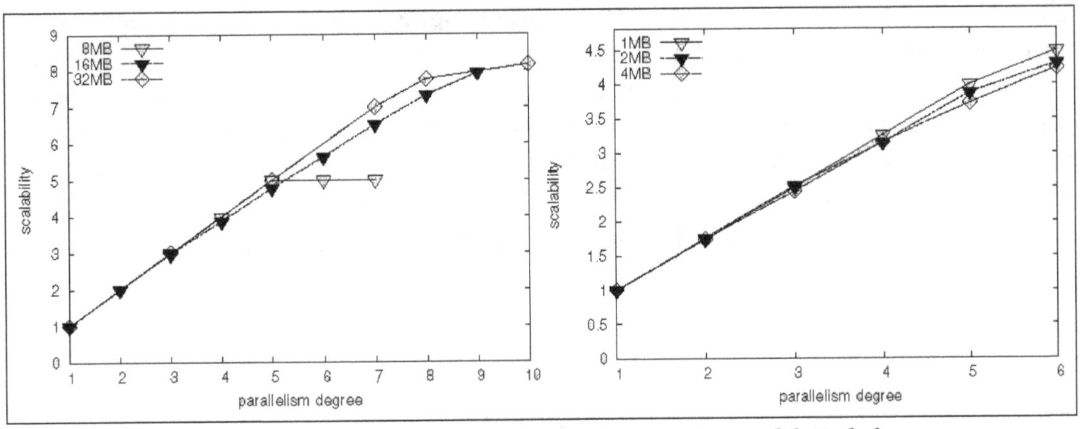

Scalability of different versions of the flood forecasting model: Task-farm on a cluster (left), data-parallel on a multicore architecture (right).

The two versions of the Solver ParMod are suitable for the execution on different computing architectures. The task-farm structure does not require intensive data exchanges between processes, therefore it can be efficiently implemented on a distributed-memory architecture (e.g. a cluster of production workstations). The data-parallel approach is instead amenable of being executed on a shared-memory architecture as a multicore platform. Scalability of the two versions is depicted in figure, showing how different parallel versions achieve near optimal scalability results on highly different computing architectures.

Since the task-farm is based on replication of input data of different stream elements whereas the data-parallel applies a partitioning of the input data of the same task, we expect a different memory utilization of the two parallelization approaches. As described in, models describing the actual memory utilization of a structured parallel computation in function of its configuration parameters, as the parallelism degree and the parallelism scheme, provide accurate results as we can observe in the experiments depicted in figure.

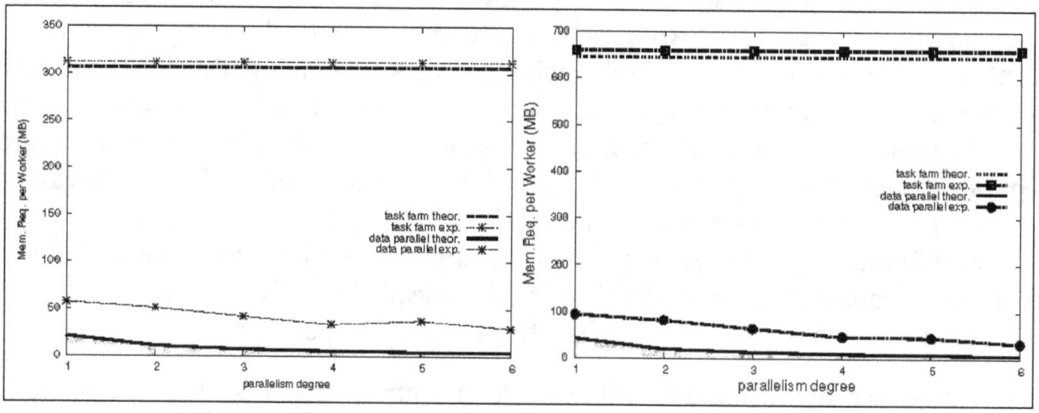

Memory utilization per worker with tasks of size 16 MB (left) and
32 MB (right): Comparison between task-farm and data-parallel schemes.

The experienced memory behavior follows the predicted values: i.e. for the task-farm scheme, based on a pure replication, the worker memory usage is independent on the parallelism degree; for the data-parallel scheme, based on a partitioning of input data, increasing the parallelism degree results on smaller partitions and thus on a lower memory usage per worker. This demonstrates how a proper versioning of a parallel module is useful not only for exploiting in the best way as possible different architectures on which the computation can be executed, but also the memory utilization of a parallel program can be adapted in response to the current memory availability of the underlying execution platform.

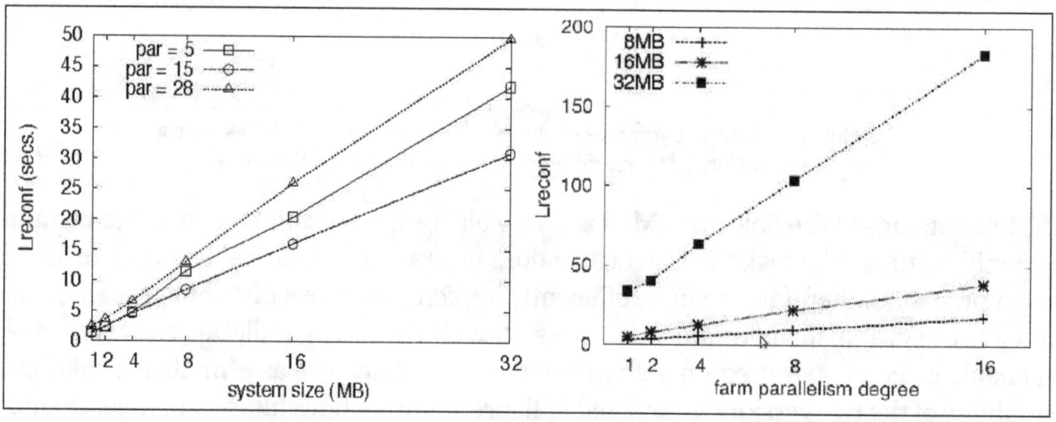

$L_{recon\,f}$ for the rollforward (left) and the rollback (right) switching form a source version (a task-farm) to a target version (a data-parallel) of the Solver ParMod. For the rollforward protocol $L_{recon\,f}$ is dominated by the time spent for completing all the pending tasks in the source version, before starting the target one. For the rollback protocol $L_{recon\ f}$ indicates the time necessary for re-executing all the rolled-back tasks.

We have discussed adaptation strategies that modify the Solver behavior in response to the actual availability of network and computing resources. Especially for functional reconfigurations, in which the ParMod computation is migrated from a source to a target environment changing the parallelism version, we need to precisely estimate the overhead induced by these activities over the execution, in order to establish if a reconfiguration action is useful for the execution or not. We have presented different reconfiguration protocols for exploiting functional reconfigurations. A rollforward protocol consists in switching from the source to the target version only when all the pending tasks in execution in the source version of the ParMod have been completely processed. On the other hand a rollback technique minimizes the intervention of the source version on the protocol: i.e. when a reconfiguration is started, the target version starts its execution immediately, re-executing possible tasks that have been completed (or they are in execution) in the source version of the ParMod. Also in this case, based on the structured parallel programming paradigm, we are able to quantify the overhead induced by different protocols and choosing the best one for each specific case. Figure depicts the time spent for applying a reconfiguration (i.e. reconfiguration latency L_{reconf}) in function of the task size.

Finally the exploitation of advanced control-theoretic techniques for controlling the Solver computation has been described in for a Cloud Computing scenario. We have analyzed the problem of executing the task-farm version of the Solver ParMod on a remote cloud architecture on which the cost of execution of a parallel computation is proportional to the number of virtualized resources (e.g. CPU) effectively used. Moreover, in order to discourage too many resource re-organizations, we have supposed the existence of a business model in which a fixed cost should be paid each time a new resource request is submitted to the cloud system.

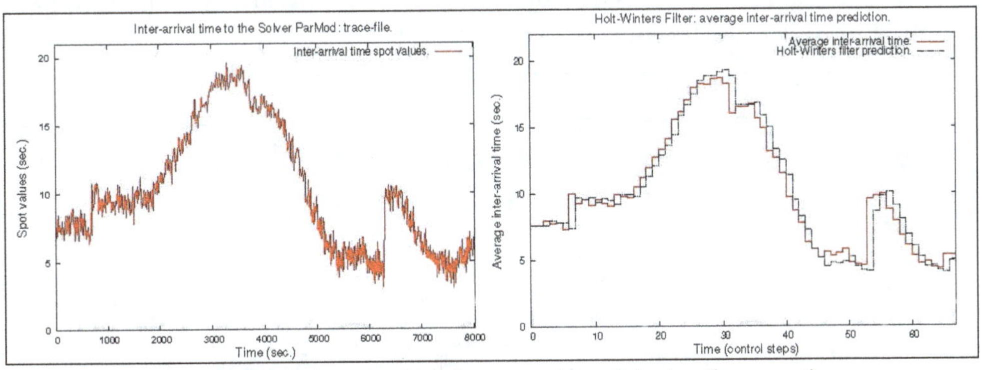

Trace-file of the inter-arrival time experienced during the execution (left) and the statistical predictions obtained with an Holt-Winters filtering (right).

In this scenario the design of effective adaptation strategies for optimizing the Solver execution are of great importance. For this application we have supposed a QoS objective with a twofold nature. First of all we need to complete the forecasting computation in the minimum completion time as possible, such that emergency stakeholders can effectively plan proper response actions in advance to potentially dangerous events. Second we need to minimize the operational cost of the Solver execution on the cloud environment.

For this reason we have adopted the model-based predictive control strategy for the Solver ParMod, starting from a hybrid modeling of interesting QoS measurements of this computation (e.g. number of completed tasks and current operational cost). Due to a time-varying network availability, we have assumed a dynamic workload consisting in a variable mean inter-arrival time 2 of tasks from the Generator module. In figure is depicted the inter-arrival time of tasks experienced during an execution scenario in which the network that interconnects the Generator and the Solver ParMods alternates congestion phases and high-reliability phases. In our modeling we have supposed the mean inter-arrival time as a disturbance input that needs to be predicted over a limited horizon. To do this we have exploited a Holt-Winters filter which is able to capture non-stationarity processes (e.g. trends and level shifts) of the underlying time-series of past inter-arrival time observations, achieving good prediction results as shown in figure.

The task-farm scheme is a parallelization pattern featuring a clear and well-defined performance model: i.e. fixing the task-farm configuration in terms of mean inter-arrival time, service times of the emitter, the collector functionalities, and the number of parallel workers, we are able to analyze the task-farm performance behavior as a queueing network and evaluating the mean inter-departure time of results from this parallel computation.

This performance modeling can be exploited in order to predict how the number of computed tasks evolves during the execution. In the MPC approach has been applied providing an utility function that describes a proper trade-off between completed tasks and operational cost. Limited prediction horizons (e.g. 1, 2 and 3 control steps) have been applied for this experiment.

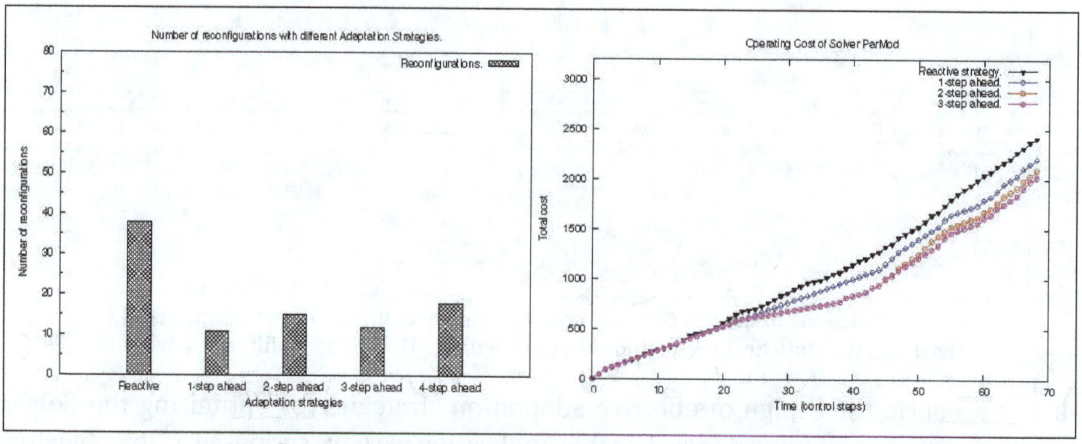

Number of parallelism degree variations (left) and long-term operational cost of the Solver ParMod execution (right).

Image (left) compares different adaptation strategies: (i) a reactive approach as the one described in, in which the parallelism degree is modified according to the actual performance measurements of the computation; (ii) a model-based predictive control technique that exploits the QoS predictability of the task-farm scheme. As we can observe the predictive approach has a positive impact on the stability degree of a ParMod configuration: i.e. for every prediction horizon length, the MPC strategy always features a lower number of reconfigurations than the reactive approach.

Image (right) depicts the long-term operating cost throughout the execution. The importance of having a predictive run-time parallelism degree adaptation is clearly highlighted. W.r.t a purely reactive adaptation, the MPC strategy is able to reduce the operating cost of even the 20% if we exploit a prediction horizon of three steps, thus demonstrating how taking reconfigurations in advance to future workload predictions can be an effective adaptation technique.

Emergency Management Information System

Emergency Management Information System (EMIS) is a computer database for disaster response that provides graphical, real-time information to responders.

EMIS and Emergencies

Four phases of an emergency are Readiness, Risk Mitigation, Response, and Replay. An EMIS shall enable emergency managers or any emergency stakeholder (affected civilians, police, fireman, Non Government Organizations (NGO), etc.) make their required activities in any phase of an emergency in an easy and speedy way.

EMIS for Readiness

- Preparation of contingency plans for different types of emergencies.
- Creating checklists that can be easily reached by any related emergency management stakeholder.
- Resource management.

EMIS for Risk Mitigation

- Determine possible risk areas and risk types. Often supported by a geographical information system (GIS).

EMIS for Response

- Executing and tracking the contingency plan.

EMIS for Replay

- Review the events of the emergency.
- Various kind of reports (supported by tables, graphs, etc).

An EMIS interacts with many other early alert systems and communicates with many legacy systems.

Advanced Disaster Management Simulator

The ADMS (advanced disaster management simulator) is an emergency and disaster management training simulation system designed to train incident commanders, first responders, and incident command teams in a real-time, interactive virtual reality (VR) environment.

Technology

The ADMS relies on a physics engine and built in artificial intelligence to provide realistic, 3D emergency situations. The Disaster scenarios include algorithms which take into account: type of threat, time of day, precipitation, wind, visibility, condition of casualties, terrain, and traffic and bystander behavior ADMS training exercises are unscripted and open-ended, requiring interactive decision making and participation from the users to affect the outcome of the training exercise. The development, escalation, or resolution of the situation is determined by the trainees' decisions and the use of intelligent resources they command.

ADMS is a networked framework and a family of applications which may be run on a single station or as a multi-user system with several networked stations giving independent access to several viewpoints and control interfaces within the same environment and scenario. ADMS projects its simulations using panoramic high-definition multimedia interface, visual displays and directional sound. The simulation engine includes key models including: artificial intelligence, physics, logic, and messaging and applications comprising a visual engine, sound engine, messaging daemon and graphical user interface.

Environments

ADMS environments are developed in either geo-typical or geo-specific environments. Geo-specific environments are created using exact 3D modeling of the specified location, and can include buildings, streets, vehicles, terrain and people, specific airports or schools, or entire cities.

Products and Applications

- ADMS-COMMAND: Designed for training incident commanders in a diverse range of emergency scenarios, ranging from car accidents to low-frequency, high casualty situations, either in single or multi-agency operations. Trainees make decisions and give verbal commands, enabling the real-time action of the simulator. The system can be expanded to include additional simulation elements, including vehicle controls and cabs, and 180 degree immersive visual displays.

- ADMS-DRIVE: Designed to train, assess and recertify drivers in safe and effective driving procedures. It is also utilized as a training platform to train emergency responders to drive under stressful situations, and by airport snow removal teams to train in snow removal operations during winter conditions. ADMS-DRIVE immerses the user in a realistic, virtual environment in which dynamic elements such as traffic, signal lights, people and weather create a realistic and challenging representation of the situation.
- ADMS-ARFF: Designed as an aircraft rescue and firefighting vehicle simulator which focuses on correct turret operation, driving and communication, vehicle positioning, firefighting with turrets and Command and Control. The trainee uses controls and joysticks to maneuver the vehicle in an airport environment and deals with airplane disaster.

Permissions

All chapters in this book are published with permission under the Creative Commons Attribution Share Alike License or equivalent. Every chapter published in this book has been scrutinized by our experts. Their significance has been extensively debated. The topics covered herein carry significant information for a comprehensive understanding. They may even be implemented as practical applications or may be referred to as a beginning point for further studies.

We would like to thank the editorial team for lending their expertise to make the book truly unique. They have played a crucial role in the development of this book. Without their invaluable contributions this book wouldn't have been possible. They have made vital efforts to compile up to date information on the varied aspects of this subject to make this book a valuable addition to the collection of many professionals and students.

This book was conceptualized with the vision of imparting up-to-date and integrated information in this field. To ensure the same, a matchless editorial board was set up. Every individual on the board went through rigorous rounds of assessment to prove their worth. After which they invested a large part of their time researching and compiling the most relevant data for our readers.

The editorial board has been involved in producing this book since its inception. They have spent rigorous hours researching and exploring the diverse topics which have resulted in the successful publishing of this book. They have passed on their knowledge of decades through this book. To expedite this challenging task, the publisher supported the team at every step. A small team of assistant editors was also appointed to further simplify the editing procedure and attain best results for the readers.

Apart from the editorial board, the designing team has also invested a significant amount of their time in understanding the subject and creating the most relevant covers. They scrutinized every image to scout for the most suitable representation of the subject and create an appropriate cover for the book.

The publishing team has been an ardent support to the editorial, designing and production team. Their endless efforts to recruit the best for this project, has resulted in the accomplishment of this book. They are a veteran in the field of academics and their pool of knowledge is as vast as their experience in printing. Their expertise and guidance has proved useful at every step. Their uncompromising quality standards have made this book an exceptional effort. Their encouragement from time to time has been an inspiration for everyone.

The publisher and the editorial board hope that this book will prove to be a valuable piece of knowledge for students, practitioners and scholars across the globe.

Index

A

Advanced Disaster Management Simulator, 238

Alpide Belt, 44

Arctic Freeze, 113-114

Artificial Induction, 46

B

Breaking Wave, 58, 63

C

Capacity Assessment, 161, 164

Channel Slope, 53

Chemical Cloud, 108-109

Climate Change, 56, 91

Community-led Approach, 117

Cost-benefit Analysis, 12-13, 125

Crisis Communications, 16-18, 89

Critical Infrastructure, 86, 107, 114

D

Damage Assessment, 108, 114, 137-139, 143

Debris Flow, 10, 78-79, 81, 83

Disaster Cycle, 92, 105

Disaster Management, 5, 87, 89, 104, 117, 151, 158, 161, 181-182

Disaster Mitigation, 8, 55, 67, 85, 120, 144-146, 151, 159, 168, 179, 182

Disaster Recovery Plan, 49, 124-129, 133-136, 138-141, 143

Disaster Response, 85, 87-90, 104-106, 139, 151, 237

Disaster Risk Reduction, 41, 93, 164, 182

Disease Epidemics, 42

E

Earthquake Preparedness, 48

Elastic Strain, 44, 46, 206

Emergency Alert System, 18-19, 67, 109, 179, 192, 194, 205

Emergency Personnel, 30, 105

Emergency Population Warning, 179, 191-192

Emergency Preparedness Kit, 22, 30, 40

Emergency Response Plan, 50, 89, 94, 107, 112, 114, 116, 188

F

Fire Alarm System, 14, 108, 179, 183-190

Fire Sprinklers, 19, 189

First Aid Kit, 40, 75, 100, 102

Flash Floods, 1, 22, 26, 193

Flood Forecasting, 56-57, 232-233

Food Handling, 35, 40

G

Gale Warning, 179, 209

Generic Preparedness, 86, 88, 97

Geographic Information Systems, 10, 96, 159

Geomorphological Hazards, 9

H

Hazard Management, 150-151, 154-155, 158-159

Hazard Vulnerability Analysis, 99

Hospital Safety Index, 99

Humanitarian Aid, 40-41

Hurricanes, 1, 28, 72, 74, 115, 123, 144, 154, 156, 192

I

Incident Management, 18, 93, 210

Indicator-based Approach, 12-13

Information Security Management, 123

Intensity Scales, 64

M

Man-made Disasters, 2, 41, 105

Material Safety Data Sheets, 111

Mock Drills, 128, 143

N

Natural Hazards, 10, 66, 112, 123, 144, 147, 151-157, 159, 175-178, 182

Needs Assessment, 14-15, 85

P

Performance Objectives, 14-15, 107, 110, 112
Personal Protective Equipment, 17, 88, 97
Public Preparedness, 90

Q

Quality Assurance, 96-97

R

Radiation Disaster, 91
Reconstitution Phase, 136, 141-142
Recovery Point Objective, 121-122, 125
Rehabilitation Costs, 180
Resource Allocation, 85, 95, 106
Risk Assessments, 10, 86, 153, 157
Risk Mapping, 10, 12-13

S

Seismic Data, 204
Seismic Waves, 42, 44-45, 48, 202, 209
Seismicity Maps, 43
Storm Surge, 114, 162, 198
Subduction Zone, 206
Survival Kit, 40

T

Tectonic Earthquakes, 44-45, 47, 59
Tectonic Plates, 42-43
Thunderstorm, 21, 24-27, 41, 71, 109, 114, 132, 194
Thunderstorm Risk, 24
Tidal Wave, 58, 162
Tornado, 16, 19, 21, 24, 28-30, 41, 68-76, 108-109, 113-114, 132, 194, 198-199
Tornado Drills, 75-76
Train Personnel, 112
Tsunami Heights, 65
Tsunami Warning System, 66-67, 179, 201-204

V

Volcanic Eruptions, 58-59, 78, 175, 192, 203
Volcanism, 46

W

Whole Community Activity, 103